AF358621

Sustainable Management of Urban Water Resources

Sustainable Management of Urban Water Resources

Editors

Susanne Charlesworth
Craig Lashford

MDPI • Basel • Beijing • Wuhan • Barcelona • Belgrade • Manchester • Tokyo • Cluj • Tianjin

Editors
Susanne Charlesworth
Centre for Agroecology,
Water and Resilience,
Coventry University
UK

Craig Lashford
Centre for Agroecology,
Water and Resilience,
Coventry University
UK

Editorial Office
MDPI
St. Alban-Anlage 66
4052 Basel, Switzerland

This is a reprint of articles from the Special Issue published online in the open access journal *Water* (ISSN 2073-4441) (available at: https://www.mdpi.com/journal/water/special_issues/Management_Urban_Water).

For citation purposes, cite each article independently as indicated on the article page online and as indicated below:

LastName, A.A.; LastName, B.B.; LastName, C.C. Article Title. *Journal Name* **Year**, *Volume Number*, Page Range.

ISBN 978-3-03943-893-8 (Hbk)
ISBN 978-3-03943-894-5 (PDF)

Cover image courtesy of Craig Lashford.

Contents

About the Editors

Susanne Charlesworth is Professor of Urban Physical Geography at Coventry University in the Research Centre for Agroecology, Water and Resilience. She is the author of more than 80 peer-reviewed journal articles on urban pollution and sustainable drainage systems (SuDS) and many book chapters, and she has co-edited books on aquatic sedimentology, water resources, and SuDS. She is particularly interested in the application of SuDS to challenging environments such as refugee camps and informal settlements.

Craig Lashford is an Assistant Professor in Physical Geography at Coventry University in the Research Centre for Agroecology, Water and Resilience. His research focuses on modelling the impacts of different approaches to sustainable flood management, with particular interest in sustainable drainage systems (SuDS).

Preface to "Sustainable Management of Urban Water Resources"

Currently, 55% of the world's population lives in urban areas, and this figure is predicted to grow to 68% by 2050, adding more than 2.5 billion people to urban populations. The United Nations World Water Development Report, 2018, warns that by 2030, the global demand for fresh water is likely to exceed supply by 40%. Added to population growth, climate change has the potential to lead to changes in rainfall regimes, with the potential of increased flooding and drought. Currently, 1.2 billion people are at risk from flooding, but this is predicted to increase to about 1.6 billion, i.e., nearly 20% of the world population, by 2050. To address these issues, approaches are needed that are flexible and have multiple benefits. This Special Issue includes topical issues around the management of urban water from groundwater supplies, the use of modelling to assess the use of sustainable drainage management trains at the construction site scale to address urban flooding, the management of surface water using approaches based on mimicking nature at the small scale, and the issues around the impacts of urbanisation on water quality and sustainable protection of the urban coastal zone.

Susanne Charlesworth, Craig Lashford
Editors

 *water*

Article

The Groundwater Demand for Industrial Uses in Areas with Access to Drinking Publicly-Supplied Water: A Microdata Analysis

Pilar Gracia-de-Rentería [1,*], Ramón Barberán [2,3] and Jesús Mur [4]

[1] Agrifood Research and Technology Centre of Aragon (CITA), Montañana Avenue, 930, 50059 Zaragoza, Spain

[2] Department of Public Economics, Faculty of Economics and Business, University of Zaragoza, Gran Vía Street, 2, 50005 Zaragoza, Spain; barberan@unizar.es

[3] Environmental Science Institute (IUCA), University of Zaragoza, Pedro Cerbuna Street, 2, 50009 Zaragoza, Spain

[4] Department of Economic Analysis, Faculty of Economics and Business, University of Zaragoza, María de Luna Street, Campus Río Ebro, 50018 Zaragoza, Spain; jmur@unizar.es

* Correspondence: mpgracia@cita-aragon.es; Tel.: +34-976-716356

Received: 3 December 2019; Accepted: 7 January 2020; Published: 10 January 2020

Abstract: This study examines, from an economic perspective, the factors influencing the decision of companies to use groundwater or not, in a context in which they have access to drinking publicly-supplied water and can also opt for self-supplying groundwater, and then estimates its groundwater demand. The Heckman two-stage model is applied, using microdata of a sample of 2579 manufacturing and service companies located in Zaragoza (Spain). The results of the first stage show that companies have economically rational behavior in the choice of their water supply sources: the probability to capture groundwater depends negatively on its cost and positively on the cost of publicly-supplied water. The results of the second stage indicate that the demand for self-supplied groundwater is normal, but inelastic (elasticity of −0.50), and that self-supplied and publicly-supplied water are substitutive inputs, where the cross-elasticity of the demand is much higher than the direct elasticity. These results warn of the undesirable consequences, on overall efficiency and environmental sustainability, of the lack of a volumetric fee that charges companies with the environmental and resource costs caused by the extraction of groundwater and emphasize the need for integrated management of all water resources.

Keywords: groundwater; Heckman model; self-supply; water demand; water economics; industry

---

1. Introduction

Water is an essential resource for socio-economic development, human life sustenance, and ecosystem preservation. Therefore, it is necessary to ensure the sustainability of water resources and their efficient and equitable allocation to enable an acceptable level of economic and social welfare. Nevertheless, population growth, urbanization, water pollution, and unsustainable development are all increasing pressure on water resources across the world, and that pressure is further exacerbated by climate change [1]. Pressure affects both surface water and groundwater.

There is general agreement on the importance of groundwater and the severity of the pressures it bears. Groundwater comprises a much larger freshwater volume than surface water and it is increasingly important for water security in many countries and regions, but many aquifers are subject to unsustainable abstraction levels and pollution [2,3]. The United Nations [4] reports that groundwater provides drinking water to at least 50% of the global population and accounts for 43% of all water used for irrigation, and that an estimated 20% of the world's aquifers are overexploited.

The main causes of this overexploitation of aquifers are the abstraction for irrigation, drinking water, industrial and mining uses [1]. The relative importance of each of these uses varies significantly by country, depending on climate and the degree of economic development. The problem of overexploitation arises in fossil aquifers because of their lack of natural replenishment, and in aquifers with natural recharge when groundwater is withdrawn faster than its long-term replenishment. The main consequences are falling groundwater levels, increased pumping costs, land subsidence, reduced baseflows of rivers (desiccation of springs, streams, and wetlands), water quality degradation, saline intrusion, and rising sea levels [5–7].

The proposed solutions involve both an increase in water supply and its conservation: in the first case, through artificial recharge of aquifers and interventions to improve groundwater quality; in the second case, through the implementation of administrative controls and economic incentives to reduce abstractions [8]. Furthermore, given the close relationship between surface water and groundwater [9–11], the long-term sustainability of their use requires integrated management of all water resources, in line with the approach adopted in the Water Framework Directive [12] and the recommendations of the United Nations [1–3]. Unfortunately, the implementation of policies aimed at the sustainability of groundwater exploitation faces serious difficulties, as evidenced in its increasing deterioration. These problems can be mainly attributed to the invisibility of groundwater which limits the availability of information on the real situation of aquifers and also to its character as a common-pool resource in the sense of Hardin [13], which encourages users to overexploit [14].

Typically, groundwater use occurs in water-stressed regions, where aquifers are used as an additional source to surface water, but also occurs in regions without water scarcity and where the supply of water from other sources is sufficient and secure, as in urban settings in developed countries. In these urban areas, households and industries have access to the drinking water provided by the public water supply network, but they sometimes complement or replace that public supply with self-supply of groundwater when they use water for some purposes which do not require drinking water. The possibility of choosing between alternative water sources has relevant consequences for the management of aquifers and public drinking water supply services, since the measures adopted by policy makers regarding one of these sources will surely affect the other and vice versa.

Despite the key role of self-supply, this water source has been barely analyzed by the economic literature. This lack of empirical evidence is more pronounced in the case of industry (where we can only cite Ref. [15,16]) than in that of households (Ref. [17–20] among others), where there is a broader literature focused on developing countries in which the low reliability in public supply leads households to use other alternative sources (wells, rainwater tanks, public water fountains, water vended from tank trucks, bottled water). The studies regarding self-supplied water encountered problems regarding lack of information because microdata are rarely available to the public. As a consequence, researches face serious difficulties in knowing the quantity of intake water and the cost born by each user. In the case of groundwater, there is usually an absence of public meters for monitoring the water extracted by each user, and a lack of statistics on its unitary cost, which includes the costs of investment, extraction, and treatment.

The mainstream of literature that estimates water demand implicitly assumes that the source of supply is determined exogenously, both for analyzing publicly and self-supplied water. Focusing on self-supply for industrial activities, this was the case in a study by Reynaud [16] who estimated self-supplied water demand for 55 industrial and service companies located in France. However, a suitable approach should take into account that a considerable number of users can choose their water sources and how much to use from each one. The literature has usually addressed this issue by means of a two-stage process where in the first stage the user decides whether to use a given source (for example, self-supply), and in the second stage, decides on the volume of water to capture. This strategy is rather common when analyzing water recirculation [21–23] and self-supply in the domestic sphere (for example, [17–20]), but for industrial users, we can only refer to a study by Renzetti [15], who estimated self-supplied water demand in the US using a survey of more than 2000 manufacturing firms. This shows that more empirical evidence is needed on the choice and relationship between water supply sources and the estimation of groundwater demand in the industrial field, in order to establish adequate policies for an integrated water management.

The purpose of this study is to analyze the factors that influence decisions on the use of self-supplied groundwater by manufacturing and service companies in urban settings in which they can also choose to supply from the public drinking water network, and to estimate the groundwater demand for these activities. The study is based on a sample of 2579 companies located in Zaragoza (Spain), 44 of which use self-supplied water. We use the Heckman two-stage model, which allows us to obtain the marginal effect of the different factors on the probability of self-supply and on the volume of self-supplied groundwater. Our attention is focused on economic factors since, in the absence of technical impediments, it will be the expectation of benefit from water use that will induce companies to choose to pump water from the aquifer [24]. The analysis is oriented towards the design of public policies to promote sustainability and efficiency in the use of water resources.

After this introduction, Section 2 presents the case study. Section 3 describes our data. Section 4 introduces the model and the corresponding estimation techniques. The results are discussed in Section 5. Finally, Section 6 presents the main conclusions.

2. The Case Study

The municipality of Zaragoza has the fifth largest population in Spain. Its production structure is similar to the national average, characterized by the dominance of the service sector (84% of employment), followed by manufacturing (10%), construction (5%), and farming (1%), according to data for 2012 from the Aragonese Statistics Institute [25].

The municipality is located in the center of the Ebro River basin, at the mouths of two tributaries, the Gállego and the Huerva rivers. The management and administration of the different water masses in this basin are the responsibility of the Ebro Hydrographic Confederation (CHE), a public agency dependent on the Spanish government. The drinking water supply in the municipality has traditionally come from the Imperial Canal of Aragon, which runs alongside the Ebro, the source of its water, although since 2010 it has been supplemented with water channeled from the Pyrenees. The drinking water supply and wastewater services are the responsibility of the Zaragoza City Council. Both services are taxed by a binomial tariff system which combines a fixed and variable charge (volumetric charge). The fixed charge depends on the caliber of the meters which measure the water supplied to each user and the variable charge depends on the volume of water recorded in these meters and is obtained by applying an increasing block tariff.

There are two groundwater masses underlying the municipality of Zaragoza: The Ebro-Zaragoza alluvial aquifer and the river Gállego alluvial aquifer (see Figure 1). These two groundwater masses, known as the Zaragoza aquifer, provide the municipality with an abundant water source, easily accessed using wells only about twenty meters deep. The groundwater resources' availability is also common in many other areas of Spain where aquifer systems cover two thirds of the surface area [26] and, on average, groundwater meets around 20% of the demand for water, but can represent up to 75% of total water use in the Mediterranean Basin [7].

The water extracted from the Zaragoza aquifer has a constant temperature and is turbidity-free, so it does not usually need any treatment before its use for certain industrial purposes. For current extraction levels, there are no overexploitation problems, so it is a source with almost guaranteed availability [27–29]. In the Ebro River basin, the use of this resource is subject to the concession of a license by the CHE (according to the 1985 Spanish Water Act [30]), who authorizes a maximum volume of water extraction based on the request of each user. In order to control this volume, users are obligated to install private meters for their monitoring [31]. However, the lack of public homologated meters, along with the difficulty of monitoring all existing wells in the river basin, imply that the CHE does not have official records on the real volume captured by all users, but only for some specific users or for some water bodies with serious problems of water availability, which is not the case of the Zaragoza aquifer. Therefore, water extraction control in practice is mostly based on occasional inspections to verify that users do not exceed the maximum volume authorized. The direct discharge of water into water bodies is also subject to an authorization and control process similar to that of water extractions [31] and faces similar problems in its practical application.

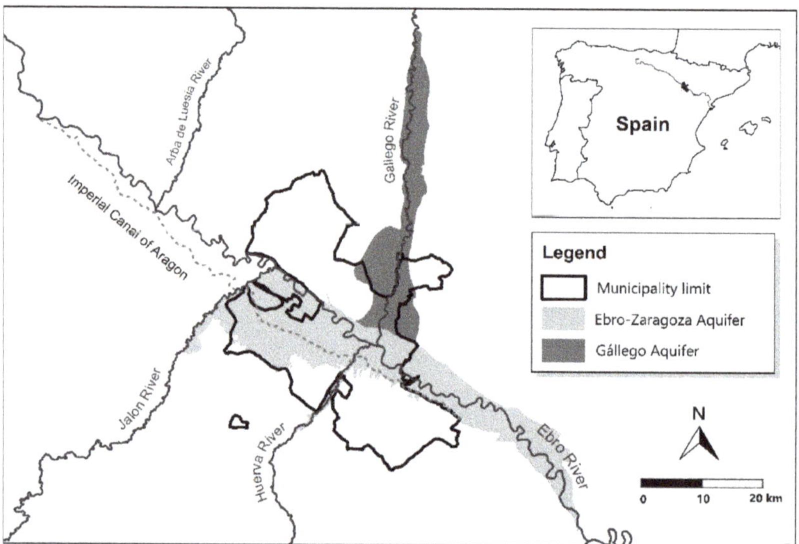

Figure 1. Location of the Ebro-Zaragoza and Gállego alluvial aquifers. Source: By the authors, based on [27,28].

Unlike publicly-supplied water, the use of this resource is not subject to a supply tariff; it is subject only to a one-off administrative fee linked to the licensing procedure for groundwater extraction and designed to cover the costs of the procedure. On the contrary, users do must pay for the discharge of this resource. If the self-supplied water is discharged directly into the river channels or into the aquifer, a dumping fee [32] should be paid to the CHE based on the volume of water discharged authorized and the quality of the discharge (this fee is very low, 0.03005 €/m^3, and can vary depending on the quality of the discharge). If it is discharged into the municipal sanitation network, the municipal wastewater tariff should be paid to the City Council (this tariff includes both the dumping fee plus the corresponding wastewater treatment costs). This means that the unit cost of self-supply born by users (including the license fee, the cost of groundwater extraction, well drilling, pumping equipment and pumping water, and the cost of discharge) is, in most cases, lower than the publicly-supplied water tariff.

This easy accessibility has led to strong pressures on these water bodies in terms of quality [27,28]. This implies that these aquifers are in risk of not achieving the good qualitative status of water bodies established in the Water Framework Directive as the 30.9% of Spanish groundwater masses [33]. The origin of these pressures depends on the uses of water. In the case of Zaragoza, 92% of groundwater extractions are intended for the industrial sector [29], which also represents an important source of pollutants. However, for the whole of Spain the main groundwater withdrawer is irrigated agriculture, which accounts for 75% of the extractions [7].

3. Data

We have a sample of 2579 companies located in Zaragoza over the aquifer. For each company we observed the following data in 2012: the volume of publicly-supplied water and its fixed and variable cost, obtained from data provided by the Zaragoza City Council; the volume of self-supplied groundwater, obtained by combining information from the City Council and the CHE; the fixed and variable cost of self-supply, calculated from data provided by the CHE; and the value of production and the sector of activity, from the database "Iberian Balance Sheet Analysis System" (SABI) (http://informa.es/en/financial-solutions/sabi).

The data on the volume of publicly-supplied water were obtained based on records of the water meters installed in each company by the municipal water service. The data on the volume of self-supplied groundwater, in the absence of public meters, were calculated by means of two complementary procedures: the first one, as a difference between the volume discharged into the municipal sanitation network and the volume captured from the municipal supply network, based on information from the meters installed by the municipal water service; the second one, as the volume authorized in the license to use groundwater, based on records from the CHE. With the data from the City Council, we monitored the companies discharging used self-supplied water into the municipal sanitation network, and with data from the CHE we monitored the companies that instead discharge it into river channels or into the aquifer itself. Using both types of information, it was possible to build a dataset of companies who obtained water through self-supply, since no company in our sample uses surface water for self-supply, according to the CHE (this is mainly due to the poor quality of this source of water and its reduced flow in many months of the year).

To calculate the cost of self-supplied groundwater, we first need information on the depth of the aquifer at the location of the company and on the flow rate of the self-supplied water. From the geographical coordinates for each company, taken from SABI, the Geological and Mining Institute of Spain (IGME), in collaboration with the CHE, provided us information on aquifer depth, based on IGME [34] and Moreno et al. [29]. Table 1 shows the average aquifer depth for our sample (19.82 m) of self-supplying companies located in areas where the aquifer has a lower depth (16.50 m) compared to companies who do not self-supply (19.87 m). The flow rate of each company's self-supplied water was estimated based on their volume of self-supplied water, assuming that they pump 16 h a day, according to the standard of the Spanish Ministry of Agriculture, Food and the Environment [35] for water captured by industries.

Table 1. Main magnitudes relating to water consumption. Average per company for 2012.

	Aggregate	Manufacturing	Services
No. of companies with self-supply of groundwater	44	24	20
No. of companies without self-supply of groundwater	2535	242	2293
Percentage of companies with self-supply of groundwater (%)	1.71	9.02	0.86
For all Companies:			
Production (1000 €)	1482.89	2999.25	1308.51
Quantity of publicly-supplied water (m^3)	377.90	542.57	358.96
Quantity of self-supplied groundwater (m^3)	336.89	1066.30	253.00
Total water consumed (m^3)	714.78	1608.87	611.97
Percentage of self-supplied groundwater (%)	47.13	66.28	41.34
Quantity of self-supplied groundwater per € of production (L/€)	0.28	0.38	0.27
Quantity of publicly-supplied water per € of production (L/€)	0.80	0.30	0.86
Aquifer depth (m)	19.82	19.63	19.84
Companies with Self-Supply of Groundwater:			
Production (1000 €)	12,127.22	11,808.07	12,510.21
Quantity of publicly-supplied water (m^3)	4735.19	1619.9	8473.55
Quantity of self-supplied groundwater (m^3)	19,746.27	11,818.12	29,260.05
Total water consumed (m^3)	24,481.47	13,438.02	37,733.60
Percentage of self-supplied groundwater (%)	80.66	87.95	77.54
Quantity of self-supplied groundwater per € of production (L/€)	4.47	4.17	4.85
Quantity of publicly-supplied water per € of production (L/€)	0.71	0.18	1.35
Aquifer depth (m)	16.50	16.89	16.04
Companies without Self-Supply of Groundwater:			
Production (1000 €)	1298.14	2125.66	1210.81
Quantity of publicly-supplied water (m^3)	302.27	435.73	288.18
Quantity of publicly-supplied water per € of production (L/€)	0.80	0.31	0.85
Aquifer depth (m)	19.87	19.89	19.87

The annual fixed cost of self-supplied groundwater (FCS) was calculated according to [35], as follows:

$$FCS = F + CC + CM + OMC \tag{1}$$

where F is the one-off administrative fee that users must pay when processing the license to use groundwater, assumed to be valid for 20 years according to [36]; CC is the cost of well construction (drilling, laying pipes, and finishing the well), supposing this to be amortized over 20 years; CM is the cost of investment in machinery (pumping equipment), to be redeemed in 10 years; and OMC is operating and maintenance costs (representing 2% of the investment cost).

We calculated the well construction costs based on the depth of the aquifer, while the cost of the pumping equipment was obtained according to its market price, depending on the power needed for the pumps. The power was obtained using the approximation of [37]:

$$P = \frac{h \times Q}{r \times 75} \tag{2}$$

where P is the power (in metric horsepower); h is the manometric height (in meters), which we make equal to the aquifer depth; Q is the flow rate (in liters per second); r is pump performance, considered to be 70% in all cases ($r = 0.70$); and the constant 75 in the denominator enables us to go from kilogram-meters per second to metric horsepower.

The variable unit cost of self-supplied groundwater (VUCS) is the cost of the energy needed to capture a cubic meter of water, plus the cost of the municipal sanitation charge for companies which discharge self-supplied water into the municipal sanitation network, or the cost of the dumping fee paid to the CHE otherwise.

We calculated the energy cost per cubic meter of water extracted (UEC) according to [37], as follows:

$$UEC = 0.002726 \frac{h \times k}{r} \tag{3}$$

where h is the manometric height (in meters); k is the approximate price of energy (€/Kwh) for the average price of electricity in Spain [38]; r is pump performance (again, set at 70%); and the constant 0.002726 is energy consumption (Kwh) incurred by raising one m^3 of water one meter.

For companies that only use publicly-supplied water, we need to know the fixed and variable unit cost of self-supplied water that they would face if they decided to capture water from the aquifer. So, for these companies, we calculated FCS and VUCS supposing that, if they decided to self-supply, they would capture the same percentage of self-supplied water as the average for companies that self-supply.

The annual fixed cost of publicly-supplied water (FCP) is obtained as the annual municipal fixed charge of the publicly-supplied water supply and sanitation bill. In turn, we calculated the variable unit cost of publicly-supplied water (VUCP) by dividing the municipal variable charge of the publicly-supplied water and sanitation bill by the intake volume.

As before, we need to estimate the fixed and variable unit costs of publicly-supplied water that companies would encounter by using self-supplied water, if they decided to use only the public water network. In these cases, we estimated the corresponding FCP and VUCP assuming that, if they decided to publicly-supply, they would capture the same volume in publicly-supplied water as they do in self-supply.

Finally, SABI contains information about the value of production (Y) and the sector of activity each company belongs to (manufacturing or services). Based on the last item, we generated the corresponding dummy variable (DM).

Table 1 offers additional detail in relation to companies that use self-supplied and publicly-supplied water. A total of 1.71% of companies in our sample capture water from the aquifer, but the self-supplied water used by these firms represents 47.13% of total water used by the industrial sector. In the manufacturing sector, the percentage of companies is 9.02% (representing 66.28% of total water volume), while in the services sector the percentage of companies is lower than 1% (representing 41.34% of water use). We also observed that companies using groundwater self-supply are larger than companies using only publicly-supplied water (the average output for these groups is 12,127,220 € vs. 1,298,140 €) and use a much higher total volume of water per euro of production (5.18 L/€ vs. 0.80 L/€).

For companies using self-supply, 80.66% of the water they consume is self-supplied. Again, this percentage is higher in manufacturing companies (87.95%) than in services (77.54%). However, the volume of self-supplied water per euro of production is slightly greater in companies in the services sector (4.85 L/€) than in manufacturing (4.17 L/€).

4. Empirical Application

Our approach is based on the assumption that companies choose their sources of water (publicly and/or self-supplied water) and the amount of each in order to minimize the cost of production. This point leads us to a two-stage model where, first, the company decides whether or not to self-supply with groundwater and, if it does, then it decides the volume of water extracted from the aquifer. Section 4.1 introduces the methodological background for our approach whereas Section 4.2 discusses the application to our case study.

4.1. Methodology: Heckman Two-Stage Model

There are several alternatives to proceed with two-stage models [39]. Among the existing alternatives, we prefer the classical Heckman approach [40] because of its greater flexibility, allowing different factors to intervene in each stage. Before going into the details, we shall introduce briefly the basis of this approach.

The aim in the first stage is to model the probability that a company decides to capture groundwater, through a probit equation for a binary decision variable, h_i, such as the following:

$$h_i = 1 \ (y_i > 0) \ if \ h_i^* > 0 \atop h_i = 0 \ (y_i = 0) \ if \ h_i^* \leq 0 \quad with \ h_i^* = x'_{1i}\beta_1 + \varepsilon_{1i} \ i = 1, 2, \ldots, N \tag{4}$$

where y_i is the volume of self-supplied groundwater and h_i^* is a latent, unobserved variable representing the decision process (N is the sample size); x_{1i} is a vector of observed characteristics of the company. It is usual to assume normality for the error term of the equation, ε_{1i}. This is the decision equation, which allows us to quantify the probability of self-supply:

$$P(h_i = 1) = \Phi\left(x'_{1i}\beta_1\right) = 1 - \Phi\left(-x'_{1i}\beta_1\right) \tag{5}$$

The purpose of the second stage is to explain the volume of groundwater captured by each company, using a truncated regression model such as:

$$y_i = x'_{2i}\beta_2 + \varepsilon_{2i} \ if \ y_i > 0 \tag{6}$$

This is the quantity equation. The error terms of both equations could be correlated, $corr(\varepsilon_{1i};\varepsilon_{2i}) = \rho \neq 0$, so that the least squares estimations of the first equation would be biased. The Heckman approach corrects for this source of inconsistency introducing the inverse of the so-called Mills ratio (IMR), or non-selection hazard in Equation (6):

$$y_i = x'_{2i}\beta_2 + \rho\sigma_{\varepsilon_{1i}} \ IMR_i + \eta_i \tag{7}$$

where $IMR_i = \dfrac{\phi\left(-x'_{1i}\beta_1\right)}{1-\Phi\left(-x'_{1i}\beta_1\right)}$ with $\phi(.)$ and $\Phi(.)$ being the standard normal density and distribution functions estimated in the decision equation; x_{2i} is a vector of observed characteristics of the company, possibly different from x_{1i}. The significance of the composed coefficient, $\gamma = \rho\sigma_{\varepsilon_{1i}}$, is crucial for the specification.

Once the two-stage model is estimated, it is possible to evaluate the marginal effects. The effect of a continuous z variable on the probability of self-supply is:

$$\frac{\partial P(h_i = 1)}{\partial z} = \phi\left(x'_{1i}\beta_1\right)\frac{\partial\left(x'_{1i}\beta_1\right)}{\partial z} \tag{8}$$

The effect on the conditional volume of self-supplied groundwater is:

$$\frac{\partial E(y_i | h_i = 1)}{\partial z} = \frac{\partial\left(x'_{2i}\beta_2\right)}{\partial z} - \gamma\left(\frac{\phi\left(-x'_{1i}\beta_1\right)}{1-\Phi\left(-x'_{1i}\beta_1\right)}\right)\left[\frac{\phi\left(-x'_{1i}\beta_1\right)}{1-\Phi\left(-x'_{1i}\beta_1\right)} + x'_{1i}\beta_1\right]\frac{\partial\left(x'_{1i}\beta_1\right)}{\partial z} \tag{9}$$

Moreover, the effect of a discrete variable is the difference between the two states of the binary h_i variable.

4.2. Application to the Case Study

From Table 1, presented in the previous section, we can observe that our case study fits well with the Heckman approach; in fact, it is a two-stage decision process where the factors intervening in the two instances can vary. For example, companies in the manufacturing sector seem to be more likely to self-supply but, once they have made the decision, other factors such as volume of activity seem to be more important.

Therefore, we can adapt the Heckman model described in Section 4.1. to our case study as follows. For the first stage we have:

$$\begin{aligned} DS_i &= 1 \, if \, h_i^* > 0 \\ DS_i &= 0 \, if \, h_i^* \leq 0 \end{aligned} \quad with \quad h_i^* = x'_{1i}\beta_1 + \varepsilon_{1i} \quad i = 1, 2, \ldots, N \tag{10}$$

where:

$$x'_{1i}\beta_1 = \beta_{1,FCS}lnFCS_i + \beta_{1,VUCS}lnVUCS_i + \beta_{1,FCP}lnFCP_i + \beta_{1,VUCP}lnVUCP_i + \beta_{1,DI}DM_i \tag{11}$$

DS_i is a binary indicator of positive self-supplied groundwater. The set of k_1 first stage factors, x_{1i}, are the variables described in Section 3. Note that the variables in the right hand side of the equation have been log-transformed to be more consistent with the second stage of the procedure.

For the quantity equation of the second stage, we specify a double logarithmic model to prevent negative estimates (other functional forms were discarded based on misspecification tests), so that:

$$lnVS_i = x'_{2i}\beta_2 + \varepsilon_{2i}; \; \varepsilon_{2i} \sim N\left(0, \sigma_2^2\right) \tag{12}$$

where:

$$x'_{2i}\beta_1 = \beta_{2,1} + \beta_{2,VUCS}lnVUCS_i + \beta_{2,VUCP}lnVUCP_i + \beta_{2,Y}lnY_i + \beta_{2,DI}DM_i \tag{13}$$

VS_i is the quantity of pumped groundwater conditioned to $DS_i = 1$, and x_{2i} is a set of k_2 explicative factors ruling in the second stage (described in Section 3).

In the equation of the first stage, we include the fixed cost (investment cost) and the variable unit cost of self-supply groundwater; we expect that an increase in both variables will reduce the probability of self-supply. We also include the fixed cost and the variable unit cost of publicly-supplied water; we expect that an increase in both variables will increase the probability of self-supply. Finally, we include a sectoral dummy, for which we have not any a priori, since it would depend on the uses of water inputs in each sector. Nevertheless, data from Table 1 suggest that there is a higher percentage of self-suppling companies in the manufacturing sector (9.02%) than in the service sector (0.86%). It should be noticed that we have not included an output variable in the first stage equation. The reason is that the level of production of a company has been implicitly taken into account when including the costs of investment in the self-supply decision equation. This means that, given a certain fixed cost of self-supply, the company will decide whether this investment is profitable or not given its level of production, and therefore, whether to self-supply or not. So, the inclusion of the variable output in the equation of the first stage would have been redundant.

Regarding the quantity equation (second stage), we include the following: the variable unit cost of self-supply, for which a negative relationship with the quantity demanded of this water source is expected; the variable unit cost of publicly-supplied water, for which we did not adopt an a priori hypothesis because the sign of this relationship depends on technical factors and not just economic ones (an increase in this variable will increase the quantity of self-supply if both types of water are substitutes and reduce the quantity if they are complementary); the level of production, for which a positive relationship with the quantity demanded of this water source is expected; and a sectoral dummy. For the latter, we note again, there is not any a priori, although data from Table 1 show that service self-supplying companies seem to consume more groundwater (29,260.05 m^3) than manufacturing companies (11,818.12 m^3). We do not include in the quantity equation the fixed cost of publicly and self-supplied water. The reason is that, once the decision to self-supply is taken, and the necessary investment made, the fixed costs will not determine the amount of water that the company demands.

Table 2 presents some descriptive data of the main variables of our model; we distinguish between the first and second stage equations. We confirm that half of the companies capturing water from the aquifer belong to the manufacturing sector, while only 10% of the companies in the sample in fact belong to this sector.

Table 2. Average values of the variables for the case study (2012).

	Description	First Stage		Second Stage	
DS	= 1 if the company self-supplies; 0 if not	0.017		–	–
VS	Volume of self-supplied groundwater (m^3)	–	–	19,746.27	(60,114.48)
FCS	Fixed cost of self-supplied groundwater (€/year)	2425.88	(78,487.09)	–	–
VUCS	Variable unit cost of self-supplied groundwater (€/m^3)	0.68	(0.26)	0.85	(0.65)
FCP	Fixed cost of publicly-supplied water (€/year)	207.14	(1048.75)	–	–
VUCP	Variable unit cost of publicly-supplied water (€/m^3)	1.38	(0.59)	2.95	(0.32)
Y	Value of production (thousands of €/year)	–	–	12,127.22	(2,046,581)
DM	= 1 if the company belongs to the manufacturing sector; 0 if not	0.10		0.55	

Note: Standard deviation appears in parentheses.

The fixed cost of water captured from the aquifer (investment cost) is substantially higher (2425.88 €/year) than the fixed cost of publicly-supplied water (207.14 €/year). In addition, the variable unit cost of both sources of water is greater for self-supplying companies (0.85 €/m^3 for self-supplied water and 2.95 €/m^3 for publicly-supplied water) than for the average of the sample (0.68 €/m^3 for self-supplied water and 1.38 €/m^3 for publicly-supplied water).

Table 3 presents the results of the two-stage model estimation, obtained using Equations (10) and (11). The coefficient of the IMR, $\hat{\rho}$, is positive and significant, indicating the presence of the so-called sample selection bias. Therefore, if both equations (self-supply decision and self-supply volume) were estimated separately without entering the IMR, the estimation of the parameters of the model would be biased. This confirms the appropriateness of using the Heckman two-stage model.

Table 3. Heckman two-stage model. Results of the estimation.

	First Stage (*DS*)		Second Stage (*lnVS*)	
lnFCS	−0.6463	(0.00)	–	
lnFCP	0.1525	(0.02)	–	
lnVUCS	−0.3185	(0.00)	−0.5034	(0.00)
lnVUCP	1.9068	(0.00)	5.6738	(0.00)
DM	0.8462	(0.00)	−0.1026	(0.83)
Y	–		0.3776	(0.00)
Intercept	–		−6.1988	(0.05)

Wald test: 28.00 (0.000); $\hat{\rho} = 0.6037116$; LR test ($\rho = 0$): 16.05 (0.0001)

Note: *p*-value in parentheses; Wald test refers to the $\chi^2(4)$ that all coefficients in the quantity equation are not significant. LR test ($\rho = 0$) is the likelihood ratio test that the coefficient ρ is zero (the two equations are independent), distributed $\chi^2(1)$.

5. Discussion of Results

The results obtained in the selection equation (first stage), shown in Table 3, confirm the expected signs of the explanatory variables. Thus, an increase in the cost of investment in self-supply (FCS) or the variable unit cost of captured groundwater (VUCS) reduces the probability of self-supplying water from the aquifer. This coincides with Renzetti [15] for the case of self-supplied water and is also consistent with the results obtained in the literature for the recirculation decision [21]. Moreover, an increase in the fixed cost of access to the public supply network (FCP) or its variable unit cost (VUCP) increases the probability of self-supplying groundwater, as a way of reducing the cost of water. Once again, this is in line with Renzetti [15] for the case of self-supplied water (although in his study these variables are not significant), and with the results obtained for water recirculation by Bruneau and Renzetti [21] and Féres et al. [23]. These results confirm that the companies have an economically rational behavior in this stage of selection of the water source. These results cannot be compared with those of the literature regarding households' self-supply, because this literature focuses on developing countries with public water supply quality problems, where the decision about self-supply is dependent on the reliability of the public supply and household characteristics, but not on the costs of the different sources of water [17–20]. Also, as anticipated by descriptive data, companies in the manufacturing sector are more likely to capture water from the aquifer compared to companies in the services sector. This result may be related to the predominant uses of water in each sector. Thus, while manufacturing companies need large volumes of water for tasks which do not require high quality (cooling, washing, transporting raw materials, etc.), most service companies tend to use water only for sanitary or personal care purposes requiring drinking water. On this issue, as far as we know, the literature has not provided results to compare with ours.

The results regarding the quantity equation (second stage) show that an increase in the variable unit cost of self-supplied water (VUCS) reduces the quantity of groundwater demanded. Thus, as we expected, we obtained that the demand for groundwater is normal. Renzetti [15] and Reynaud [16] also obtained a negative sign for self-supplied water in industry for the coefficient of this variable, although it is hardly significant. However, this variable appears to be very relevant in the case of recirculated water, with a negative impact on the volume of processed water [21,22,41–43], the same as for household self-supply [17–20].

Moreover, an increase in the price of publicly-supplied water (VUCP) increases the volume of groundwater captured. Therefore, both water inputs behave as substitutes. This result differs from that obtained by Reynaud [16] according to which publicly-supplied and self-supplied water for manufacturing firms are complementary, although the elasticities obtained in this case are not significant. However, the result coincides with that usually obtained in literature focused on recirculated vs. intake water in the industry [22,23,41–44] or on the demand of different types of water by households (for example, [17–19]).

The use of self-supplied water is also positively influenced by the level of production, indicating that larger companies capture larger volumes of groundwater. This result is in line with Renzetti [15] and Reynaud [16] for the case of self-supplied water, and with the results obtained for water recirculation (for example, [24,25,41,42,44]). They also coincide with those obtained commonly in the literature focusing on industrial demand for publicly-supplied water (for example, [45–47]).

The coefficient of the dummy manufacturing variable is negative but not significant. Therefore, although manufacturing companies are more likely to capture water from the aquifer (as shown in the selection equation), once they have decided to self-supply, the sector of activity does not determine the volume of intake groundwater. In fact, most service companies only use publicly-supplied water, but those that decide to self-supply utilize groundwater for uses requiring large volumes, such as cooling or filling swimming pools.

Table 4 shows the marginal effects, obtained applying Equations (8) and (9), to the case study. The first column indicates that a 1% increase in FCS and VUCS reduces the probability of self-supply groundwater by −0.0122% and −0.0060%, respectively. In contrast, a 1% increase in FCP and VUCP increases the probability of self-supply by 0.0028% and 0.0359%, respectively. Also, the probability of self-supply is 0.0553% higher for manufacturing companies. Therefore, the most relevant effects occur through the sectoral dummy and the price of publicly-supplied water, being the magnitude of the latter effect in the middle range of values previously obtained in the literature for the probability of recirculated water, which ranges from 0.02 [23] to 0.05 [22]. For the other variables, there are no previous results in the literature for establishing a comparison.

Table 4. Marginal effects on the probability of self-supply and conditional effects of self-supplied groundwater.

	Effect on the Probability of Self-Supply Groundwater	Effect on the Conditional Volume of Self-Supplied Groundwater (Elasticities)
lnFCS	−0.0121859	–
lnFCP	0.0028746	–
lnVUCS	−0.006004	−0.5033737
lnVUCP	0.0359508	5.673772
DM	0.0552805	–
Y	–	0.377608

Note: The marginal effects of the sectoral dummy on the volume of self-supplied groundwater were not estimated, since the coefficient associated to this variable in the second stage of the model is not significant (see Table 3).

The second column reports on the effect of the variables on the volume of self-supplied groundwater for those companies that have already decided to self-supply (conditional effect). It shows that a 1% increase in VUCS reduces the self-supplied volume by −0.5034%. Therefore, demand for groundwater is inelastic. This value of the direct price elasticity is similar to that obtained for the demand of non-publicly-supplied water by households [17,20]. In the industrial sphere, it is in the middle range of those of the literature for publicly-supplied water, which ranges from −0.1 to −1.1 [45], and for recirculated water, between −0.27 [22] and −1.83 [43].

In addition, a 1% increase in VUCP raises the self-supplied volume by 5.6738%. Therefore, the demand for groundwater is highly elastic with respect to the price of publicly-supplied water. This reflects that once a company has made the necessary investments to be able to self-supply, any increase in the publicly-supplied water tariff leads to intense substitution of publicly-supplied water by self-supplied groundwater. The value of this cross-price elasticity is far superior to that obtained by previous literature, both regarding the relationship between publicly and non-publicly-supplied water demand by households, where it ranges from 0.25 [20] to 1.37 [18], and when analyzing intake water vs. recirculation, with values between 0.14 [42] and 0.52 [43].

Finally, a 1% increase in the production level (Y) raises the self-supplied volume by 0.3776%. Therefore, the increase in production has a moderate impact on the demand for groundwater. The magnitude of this output elasticity is slightly smaller than that obtained by Reynaud [16] for manufacturing self-supply (0.58). It is also below the values obtained by other studies focusing on industrial publicly-supplied water, with values from 0.71 to 1.52 [45], and water recirculation, with values ranging from 0.38 [22] to 2.4 [42].

6. Conclusions

Based on a sample of 2579 manufacturing and service companies located in the city of Zaragoza (Spain), which have access to drinking water through the public-supply network and can opt for self-supply groundwater, we analyzed the determining factors in the decision whether or not to use groundwater and then analyzed the factors influencing the decision on the amount of groundwater used. For this purpose, we applied the Heckman two-stage model. The first stage examined the decision of whether or not to capture groundwater, whereas the second stage focused on the factors conditioning the volume of self-supplied groundwater.

The results obtained in the first stage indicate that the probability to use groundwater decreases when the fixed and variable cost of self-supply increase, whereas the probability increases when the fixed and variable cost of publicly-supplied water increase and when the firm belongs to the manufacturing sector. The results of the second stage show that the demand for self-supplied groundwater decreases when its variable cost increases and increases when the variable cost of publicly-supplied water or the output level increases. All the coefficients of the explanatory variables are significant, except the sectoral dummy in the second step, and their signs are consistent indicating that companies have economically rational behavior in the use of groundwater. Of particular relevance is the result indicating that groundwater self-supply is a substitute for public supply of drinking water.

The substitutability between both types of water is good news for productive efficiency, because companies can adapt the characteristics of the water collected to their needs, and thus avoid incurring unnecessary costs. From this perspective, all those policies favoring the possibility of choice between water sources are desirable. Urban planning and land management policies can contribute in this direction through the delimitation of industrial land facilitating, whenever possible, the location of companies on places where groundwater abstraction is feasible.

However, substitutability may not be good news for global efficiency and environmental sustainability. If the unit costs reflect the true social costs of the resource, both for publicly-supplied water and for groundwater self-supplied by companies, the possibility of substitution would have a positive effect on global efficiency and would not be detrimental to sustainability; but if part of those costs is not borne by companies (as usually happens when the extraction of groundwater contributes to the overexploitation of an aquifer or its contamination and when there is water shortage), the substitution of water from the public supply network by self-supplied groundwater can cause a loss of global efficiency and contribute to environmental unsustainability. In this regard, the solution lies, at least theoretically, in introducing a volumetric fee that taxes groundwater abstraction and discharge passing on the environmental and opportunity costs of the resource to users, in line with the cost recovery principle established in the European Water Framework Directive. However, the implementation of this fee faces significant difficulties, especially due to its information requirements and the opposition of those affected in the different sectors of activity. Thus, although some countries have volumetric fees for groundwater, they hardly comply with the aforementioned characteristics [48,49].

The marginal effects of the explanatory variables are very small in the first stage (effect on the probability of self-supply groundwater) but important in the second stage (effect on the volume of self-supplied water of the companies that self-supply).

The value (−0.50) of the direct price elasticity of groundwater demand (variation of the demand for groundwater when its variable unit cost (VUCS) varies) is in the middle range of those obtained in the literature for publicly-supplied water use in industry. Since this value is conditioned by the reduced magnitude of the VUCS, the adoption of a significant volumetric fee, by increasing the amount of VUCS, is expected to increase the elasticity and, therefore, the effectiveness of the groundwater pricing as an instrument for managing demand.

The high value (5.67) of the cross-price elasticity (variation of the demand for groundwater when the variable unit cost of publicly-supplied water (VUCP) varies) is above those obtained in the literature and shows the extraordinary easiness of substituting publicly-supplied water by self-supplied water once the users have made the necessary investments to make self-supply possible. This result alerts about the overestimation of water savings induced by the increase in the price of publicly-supplied water when users have access to several water sources. Under these conditions, if policy makers do not take into account the possibility of substitution of one type of water for another, they may overstate the effectiveness of publicly-supplied water pricing as an instrument to reduce pressure on the resource. Thus, in order to achieve the sustainability objective, the need to promote integrated water management must be emphasized, in line with the Water Framework Directive's recommendation. To make progress on this matter, institutional mechanisms should be established to coordinate the different government agencies responsible for the different water masses and services associated with the water cycle, which in Spain belong to the three main government levels (city councils, autonomous communities, and the central government).

Given the limited evidence accumulated on the economic determinants of groundwater demand for industrial uses, more empirical studies are necessary. However, the results obtained in this research are in line with those obtained in other related fields, such as water recirculation in industry and water self-supply in households. The data on which this analysis is based are the best available, but they are not ideal because the lack of public meters in wells for control by water authorities forced us to estimate a proxy of groundwater use. Therefore, the systematic monitoring and control of groundwater extraction and discharge by the authorities are important conditions for improving research. In addition, in a possible future extension of this analysis, the inclusion of all alternative water inputs, such as recirculation, should be considered, although data constraints are significant.

Author Contributions: P.G.-d.-R., R.B., and J.M. conceived the methodology and made the data curation, processing and analysis, the formal analysis, the econometric estimation, and the writing of the paper. All authors read and agreed to the published version of the manuscript.

Funding: This research was funded by Spanish Government's Ministry of Economy and Competitiveness (grant number ECO2015-6578-P), and the Government of Aragón and the European Regional Development Fund, through the research groups "Public Economics" (project number S23_17R), and "GAEC" (project number S37_17R).

Acknowledgments: The authors are grateful for the support received from the Zaragoza City Council through a collaboration agreement with the University of Zaragoza in relation to research concerning water.

Conflicts of Interest: The authors declare no conflicts of interest.

References

1. UN Environment. *Global Environment Outlook—GEO-6: Summary for Policymakers*; United Nations Environment Programme (UNEP): Nairobi, Kenya, 2019.
2. FAO (Food and Agriculture Organization). *Shared Global Vision for Groundwater Governance 2030 and a Call-For-Action*; FAO Publications: Rome, Italy, 2016.
3. UN Environment. *Global Environment Outlook—GEO-6: Healthy Planet, Healthy People*; United Nations Environment Programme (UNEP): Nairobi, Kenya, 2009.
4. WWAP (United Nations World Water Assessment Programme). *The United Nations World Water Development Report 2015: Water for a Sustainable World*; UNESCO: Paris, France, 2015.
5. Wada, Y.; van Beek, L.P.H.; van Kempen, C.M.; Reckman, J.W.T.M.; Vasak, S.; Bierkens, M.F.P. Global depletion of groundwater resources. *Geophys. Res. Lett.* **2010**, *37*, L20402. [CrossRef]
6. De Graaf, I.; Van Beek, R.; Gleeson, T.P.; Sutanudjaja, E.; Wada, Y.; Bierkens, M.F.P. How Sustainable is Groundwater Abstraction? A Global Assessment. In Proceedings of the AGU Fall Meeting, New Orleans, LA, USA, 11–15 December 2017.
7. Martínez-Santos, P.; Castaño-Castaño, S.; Hernández-Espriú, A. Revisiting groundwater overdraft based on the experience of the Mancha Occidental Aquifer, Spain. *Hydrogeol. J.* **2018**, *26*, 1083–1097. [CrossRef]
8. Konikow, L.F.; Kendy, E. Groundwater depletion: A global problem. *Hydrogeol. J.* **2005**, *13*, 317–320. [CrossRef]
9. Brunke, M.; Gonser, T. The ecological significance of Exchange processes between rivers and groundwater. *Freshw. Biol.* **1997**, *37*, 1–33. [CrossRef]
10. Conant, B., Jr.; Robinson, C.E.; Hilton, M.J.; Russell, H.A.J. A framework for conceptualizing groundwater-surface water interactions and identifying potential impacts on water quality, water quantity, and ecosystems. *J. Hydrol.* **2019**, *574*, 609–627. [CrossRef]
11. Sophocleous, M. Interactions between groundwater and surface water: The state of the science. *Hydrogeol. J.* **2002**, *10*, 52–67. [CrossRef]
12. European Commission. Directive 2000/60/EC of the European Parliament and the Council of 23 October 2000 establishing a framework for Community action in the field of water policy. *Off. J. Eur. Communities L* **2000**, *327*, 12–22.
13. Hardin, G. The tragedy of the commons. *Science* **1968**, *162*, 1243–1248.
14. Shalsi, S.; Ordens, C.M.; Curtis, A.; Simmons, C.T. Can collective action address the "tragedy of the commons" in groundwater management? Insights from an Australian case study. *Hydrogeol. J.* **2019**, *27*, 2471–2483. [CrossRef]
15. Renzetti, S. Examining the differences in self- and publicly supplied firms' water demands. *Land Econ.* **1993**, *69*, 181–188. [CrossRef]

16. Reynaud, A. An econometric estimation of industrial water demand in France. *Environ. Resour. Econ.* **2003**, *25*, 213–232. [CrossRef]

17. Cheesman, J.; Bennett, J.; Son, T.V.H. Estimating household water demand using revealed and contingent behaviors: Evidence from Vietnam. *Water Resour. Res.* **2008**, *44*, W11428. [CrossRef]

18. Coulibaly, L.; Jakus, P.M.; Keith, J.E. Modeling water demand when households have multiple sources of water. *Water Resour. Res.* **2014**, *50*, 6002–6014. [CrossRef]

19. Nauges, C.; Strand, J. Estimation of non-tap water demand in Central American cities. *Resour. Energy Econ.* **2007**, *29*, 165–182. [CrossRef]

20. Nauges, C.; van der Berg, C. Demand for piped and non-piped water supply services: Evidence from Southwest Sri Lanka. *Environ. Resour. Econ.* **2009**, *42*, 535–549. [CrossRef]

21. Bruneau, J.; Renzetti, S. A panel study of water recirculation in manufacturing plants. *Can. Water Resour. J. Rev. Can. Ressour. Hydr.* **2014**, *39*, 384–394. [CrossRef]

22. Bruneau, J.; Renzetti, S.; Villeneuve, M. Manufacturing firms' demand for water recirculation. *Can. J. Agric. Econ.* **2010**, *58*, 515–530. [CrossRef]

23. Féres, J.; Reynaud, A.; Thomas, A. Water reuse in Brazilian manufacturing firms. *Appl. Econ.* **2012**, *44*, 1417–1427. [CrossRef]

24. van der Gun, J.; Lipponen, A. Reconciling groundwater storage depletion due to pumping with sustainability. *Sustainability* **2010**, *2*, 3418–3435. [CrossRef]

25. Instituto Aragonés de Estadística (IAEST). Estadística Local Zaragoza 2015. Available online: https://www.aragon.es/-/estadistica-local (accessed on 28 November 2019).

26. MMA. *Síntesis de la Información Remitida por España para dar Cumplimiento a los Artículos 5 y 6 de la Directiva Marco del Agua, en Materia de Aguas Subterráneas*; Dirección General del Agua, Ministerio de Medio Ambiente: Madrid, Spain, 2006.

27. CHE (Confederación Hidrográfica del Ebro). Aluvial del Ebro-Zaragoza. Masas de Agua Subterráneas. Available online: Ftp://ftp.chebro.es/Hidrogeologia/FichasMasas/058%20Aluvial%20Ebro%20Zaragoza.pdf (accessed on 28 November 2019).

28. CHE (Confederación Hidrográfica del Ebro). Aluvial del Gállego. Masas de Agua Subterráneas. Available online: Ftp://ftp.chebro.es/Hidrogeologia/FichasMasas/057%20G%C3%A1llego.pdf (accessed on 28 November 2019).

29. Moreno, L.; Garrido, E.; Azcón, A.; Durán, J. *Hidrogeología Urbana de Zaragoza*; Instituto Geológico y Minero de España: Madrid, Spain, 2008.

30. España. Ley 29/1985, de 2 de agosto. *Boletín Oficial del Estado*, 8 August 1985; 25123–25135.

31. España. Orden ARM/1312/2009, de 20 de mayo. *Boletín Oficial del Estado*, 27 May 2009; 43940–43966.

32. España. Real Decreto Legislativo 1/2001, de 20 de julio. *Boletín Oficial del Estado*, 24 July 2001; 26791–26817.

33. EEA (European Environmental Agency). *European Waters. Assessment of Status and Pressures 2018. EEA Report No. 7/2018*; European Environmental Agency: Luxembourg, Germany, 2018; Available online: https://www.eea.europa.eu/publications/state-of-water (accessed on 28 November 2019).

34. IGME. *Trabajos Técnicos para la Aplicación de la Directiva Marco del Agua en Materia de Aguas Subterráneas. Caracterización Adicional de la Masa de Agua Subterránea del Aluvial del Ebro-Zaragoza*; Dirección General del Agua, Instituto Geológico y Minero de España: Madrid, Spain, 2005.

35. MAGRAMA. *Realización de las Tareas Correspondientes al Proceso de P.H., Preparación, Realización y Publicación de Borradores de Planes de Gestión de Cuenca y de Definición del Programa de Medidas en la Cuenca Intercomunitaria H. Ebro*; Ministerio de Agricultura, Alimentación y Medio Ambiente: Madrid, Spain, 2009.

36. España. Decreto 140/1960 de 4 de febrero de 1960. *Boletín Oficial del Estado*, 5 February 1960; 1456–1458.

37. Custodio, E.; Llamas, M.R. *Hidrología Subterránea*; Ediciones Omega: Barcelona, Spain, 1983.

38. Eurostat Energy Data 2016. Available online: http://ec.europa.eu/eurostat/web/energy/data (accessed on 28 November 2019).

39. Wooldridge, J. *Econometric Analysis of Cross-Section and Panel Data*; MIT Press: Cambridge, UK, 2001.

40. Heckman, J.J. Sample selection bias as a specification error. *Econometrica* **1979**, *47*, 153–161. [CrossRef]

41. Dupont, D.P.; Renzetti, S. Water use in the Canadian food processing industry. *Can. J. Agric. Econ.* **1998**, *46*, 83–92. [CrossRef]

42. Renzetti, S. An econometric study of industrial water demands in British Columbia, Canada. *Water Resour. Res.* **1988**, *24*, 1569–1573. [CrossRef]

43. Renzetti, S. Estimating the structure of industrial water demands: The case of Canadian manufacturing. *Land Econ.* **1992**, *68*, 396–404. [CrossRef]

44. Dupont, D.P.; Renzetti, S. The role of water in manufacturing. *Environ. Resour. Econ.* **2001**, *18*, 411–432. [CrossRef]

45. Gracia-de-Rentería, P.; Barberán, R.; Mur, J. Urban water demand for industrial uses in Spain. *Urban Water J.* **2019**, *16*, 114–124. [CrossRef]

46. Renzetti, S. Commercial and industrial water demands. In *The Economics of Water Demand*; Renzetti, S., Ed.; Kluwer Academic: London, UK, 2002; pp. 35–49.

47. Worthington, A. Commercial and industrial water demand estimation: Theoretical and methodological guidelines for applied economics research. *Estudios de Economía Aplicada* **2010**, *28*, 237–258.

48. OECD. *Pricing Water Resources and Water and Sanitation Services*; OECD Publishing: Paris, France, 2010.

49. OECD. *Groundwater Allocation: Managing Growing Pressures on Quantity and Quality*; OECD Publishing: Paris, France, 2017.

Article

Assessment of Causes and Effects of Groundwater Level Change in an Urban Area (Warsaw, Poland)

Ewa Krogulec, Jerzy J. Małecki, Dorota Porowska * and Anna Wojdalska

Faculty of Geology, University of Warsaw, Żwirki i Wigury 93, 02-089 Warsaw, Poland;
ewa.krogulec@uw.edu.pl (E.K.); jerzy.malecki@uw.edu.pl (J.J.M.); a.b.wojdalska@uw.edu.pl (A.W.)
* Correspondence: dorotap@uw.edu.pl

Received: 14 August 2020; Accepted: 3 November 2020; Published: 5 November 2020

Abstract: Monitoring the data of groundwater level in long-term measurement series has allowed for assessment of the impact of natural and anthropogenic factors on groundwater recharge. It allows for assessing the actual groundwater quantity, which constitutes the basis for balanced and sustainable groundwater planning and management in an urban area. Groundwater levels in three aquifers were studied: the shallow and deeper Quaternary aquifers and the Oligocene aquifer in Warsaw (Poland). Statistical analysis was performed on a 27-year (1993–2019) cycle of daily measurements of groundwater levels. The studies focused on determining the range and causes of groundwater level changes in urban-area aquifers. The groundwater table position in the Quaternary aquifer pointed to variable long-term recharge and allowed for the identification of homogenous intervals with identification of water table fluctuation trends. A decrease in the water table was observed within the Quaternary aquifers. The Oligocene aquifer displayed an opposite trend.

Keywords: potable supplies; groundwater level changes; infiltration; recharge; climate changes; water efficiency

---

1. Introduction

Urban development, resulting in changes in spatial management, drought, and extreme flood events caused by climate change, significantly influences hydrogeological conditions and water efficiency [1,2]. Changes in spatial management directly influence groundwater recharge, including the infiltration, lateral inflow, surface runoff, evapotranspiration, and other elements of groundwater balance [3,4]. Urban development, therefore, results in serious problems in many areas, e.g., seawater intrusion beneath coastal cities [5,6], changes in groundwater recharge and discharge [7], and groundwater pollution [8–10].

A separate problem is the water supply in urban areas in the context of changing hydrogeological conditions. During the last 50 years, the water demand in Europe has risen gradually, which is related to the increase in the human population. Around 55% of the world's human population lives in urban areas. The population living in 17 capital cities of the EU was 71.1 million in 2014, and a stable increase in population was observed in all large European cities except Athens in the last decade [11]. This has led to a general decrease in renewable groundwater resources by 24% per person in Europe [12]. An indispensable element in preventing such hazards is groundwater valorization in urban areas, with regard to both its quantity and chemical status assessment. According to the assessment rules determined by the Water Framework Directive [13], poor chemical status was observed in 9% of groundwater bodies, and, in 75% of cases, the main reason was the decrease in groundwater levels [14]. Despite the increased population and problems related to water supply in urban areas, stabilization of water abstraction has been observed in some large European cities [14]. Diversified possibilities of water supply in a particular area, local hydrogeological conditions and uneven social–economic and industry activities result in various ranges of groundwater level positions and recharge are possible in many urban areas.

In the studies of urban areas, the recharge assessment is extremely important but difficult. The usual methods of estimating recharge are available for use in urban areas as described by Lerner [15], Spalvins [16], and Schirmer [17]. Natural infiltration is modified as a result of land use, and the total balance should include the amount of water from leaks in water supply and sewage networks, which are difficult to estimate. Water loss in numerous underground water systems in Poland and the world is large, and the actual loss varies widely [18]. As much as 60% of distributed water may be lost through leakage from the distribution system [14], and losses in the water system reach, e.g., 18% in Great Britain, 30% in France, 20–34% in Spain and the Czech Republic, 30–60% in Croatia, and as much as 75% in Albania. Based on studies in 1998, the mean loss for 195 cities in Poland was 18.6% [19]. The cumulative impact of natural and anthropogenic factors, difficult to quantify individually, is reflected in the position of the groundwater table. Determining the range of dynamics of groundwater level change and their causes in short- and long-term scales is of crucial significance for social-economic activities, politics, and planning for sustainable development [1]. The influence of changes in land use and the climate on the hydrogeological conditions in urban areas depends on the location of the study area and the hydrogeological conditions [1,20]. This requires individualized studies with regard to specific known hydrogeological conditions. Methods of analyzing the impact on hydrological and hydrogeological conditions may be subdivided into three categories: experimental studies, statistical analysis, and modelling [20].

The aim of the study was to identify cumulative natural and anthropogenic causes controlling groundwater recharge and to assess their effects reflected in the groundwater table location. The cumulative impact of natural and anthropogenic factors on groundwater in an urban area and the dynamics of groundwater levels in particular aquifers have been determined by analysis of groundwater levels based on a 27-year (1993–2019) cycle of daily measurements of groundwater levels position derived from a monitoring system located in the centre of Warsaw (Poland). The analysis was focused on three aquifers: the shallow and deeper Quaternary aquifers and the Oligocene aquifer. The reliability of this analysis was assessed on standard measurements of groundwater levels in piezometers and wells, excluding the consideration of arbitrary elements or schematization of conditions for modelling studies. Direct measurements were subject to detailed statistical analysis. Assessment of the dynamics of groundwater level changes and trends of changes in various time intervals on a long-term basis allows for determining the range and variability of groundwater recharge in an urban area. Defining the variability of groundwater level changes, particularly aquifers, change trends, and range of groundwater recharge, points to the cumulative impact of factors characteristic of urban areas. The combined influence of both geogenic and anthropogenic factors shaping groundwater levels has been analyzed.

2. Background of the Study Area

2.1. Location of the Study Area

The groundwater monitoring system is located in the center of Warsaw (Poland) in the Research Station of the Faculty of Geology at the University of Warsaw. The research station is located on the crossroads of two busy streets, Żwirki i Wigury and Banacha, in the Ochota district. The district is characterized by the prevalence of multifamily buildings, similarly to most of left-bank Warsaw. In turn, single-family houses dominate right-bank Warsaw. Larger areas of undeveloped zones (park and forest complexes) are found in the urban area peripheries (Figure 1). Warsaw is located on two sides of the Vistula River and is slightly elongated along its banks (extending ca. 30 km along the N–S direction and at ca. 28 km along the W–E direction). The urban area covers an area of 517.24 km^2. The population is 1.8 million at a density of 3500 people/km^2.

Figure 1. Location of the study area.

The city is located on two sides of the Vistula River, the largest river in Poland. Two types of terraces can be distinguished along the river: a modern floodplain covering a larger surface area in the southern part of the city and older terraces with the largest surface area in the northern part of right-bank Warsaw. Moraine plateaus extend from the river to the west as the Warsaw Plain and to the east as the Wołomin Plain. The research station is located on the Warsaw Plain (Figure 1).

Groundwater recharge and drainage are variable within the urban area. In the centre of Warsaw, infiltration and its drainage are restricted by spatial management and a sewage system (stormwater drainage) typical for urban infrastructure. Precipitation from surfaces covered by non-permeable materials (roads, pavements, and roofs) is distributed by the sewage system, and thus effective infiltration is completely reduced. Groundwater recharge takes place through lateral inflow from the neighboring areas and to a large degree through water loss from the water–sewage system of the urban area [21,22]. Infiltration has a significant contribution to groundwater recharge and resource formation in other parts of the urban area.

2.2. Geology and Hydrogeological Conditions

The research station is located within a denuded post-glacial plateau located at an elevation of 108–115 m a.s.l. and incised by a glacial valley. The oldest series drilled in the Warsaw region are Upper Cretaceous strata, developed as grey marls interbedded with limestones and covered by Oligocene strata developed as sands, silts, and clays with glauconite and phosphorites. Miocene deposits include sands, silts, and clays with interbeds of lignite. The Pliocene comprises variably colored compact clays, the so-called variegated clays, and clay and sandy silts. The Pleistocene includes glacial, fluvioglacial, and ice-dammed deposits of the Podlasian, Sanian, Odranian, and Vistulian glaciations, and the Cromerian, Mazovian, and Eemian interglacials. In the vicinity of the research station, these deposits are developed as glacial tills located on silts, sands, and ice-dammed clays (Figure 2). The geological succession of the subsurface zone comprises (from the top) anthropogenic embankment deposits, sandy silts, clayey sands, glacial tills, and fine-grained and silty sands (Figure 2). At a depth of about 7.6 m b.g.l. within fine-grained and silty sands is the first Quaternary aquifer (first aquifer) with a water table monitored by piezometers. The deeper Quaternary aquifer (second aquifer) with a confined water table is located within fine-grained and poorly sorted sands at a depth interval from 26 to 46 m b.g.l. The water table drilled during the well construction at the depth of 18 m b.g.l. became stabilized at 9.95 m b.g.l. (Figure 2).

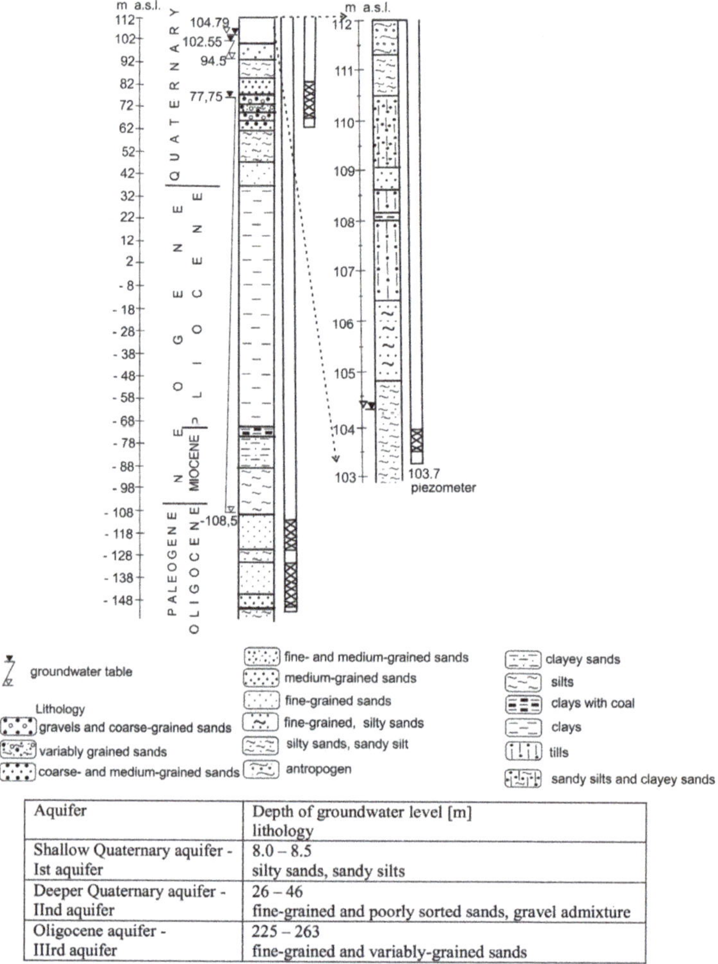

The aquifer table:

Aquifer	Depth of groundwater level [m] lithology
Shallow Quaternary aquifer - Ist aquifer	8.0 – 8.5 silty sands, sandy silts
Deeper Quaternary aquifer - IInd aquifer	26 – 46 fine-grained and poorly sorted sands, gravel admixture
Oligocene aquifer - IIIrd aquifer	225 – 263 fine-grained and variably-grained sands

Figure 2. Geological profile and location of the groundwater tables in aquifers.

Within the Warsaw agglomeration, waters of the Quaternary aquifer remain in hydraulic contact and are combined in one aquifer in the northern part of the urban area. Generally, groundwater in Quaternary deposits with hydraulic conductivity in the range of 25–30 m/day, whose thickness increases northwards and flows from the east with a velocity of 30–100 m/year [23].

Below the Quaternary aquifers is a 100-m thick series of Pliocene clays and Miocene silty sands, silts, and clays with lignite interbeds, with variable thicknesses. The deepest Oligocene aquifer (third aquifer) recognized in Warsaw in the monitoring system is found in fine- and medium-grained sands with glauconite at a depth interval from 225 to 263 m b.g.l. The groundwater table drilled at a depth of 221 m b.g.l. became stabilized at the depth of 35.75 m b.g.l. during the drilling.

2.3. Water Supply in Warsaw

The Warsaw agglomeration is supplied about 96% of its water from surface water sources, and the remaining 4% is from groundwater sources. Surface water comes from the Vistula River and the Zegrze Reservoir (located 7 km north of the Warsaw border). The water system, with a total length of 4215.7 km, covers almost the entire area of the Warsaw urban agglomeration. The population of Warsaw uses about 340,000 m^3 of water per day [24], and the daily use of water is about 135 L/person [25]. Groundwater is exploited mainly by industrial plants. About 500 intakes from the Quaternary aquifers and about 100 intakes from the Oligocene aquifer, exploiting almost 25,000 m^3/day, are located in the Warsaw agglomeration. Due to its quantity and quality, groundwater represents the strategic groundwater resources of the urban area. In the entire Mazovian voivodeship, of which Warsaw is the largest urban area, the exploitable groundwater resources from Quaternary deposits are in the range of about 205,000 m^3/h, and from the Palaeogene-Neogene deposits, they are in the range of 17,500 m^3/h [26].

3. Materials and Methods

Changes in groundwater levels in Warsaw have been studied by means of the following (Figure 3):

- statistical analysis of monitoring data from aquifers;
- comparison of groundwater level fluctuations in different aquifers;
- assessment of the trend of groundwater level changes in different aquifers;
- assessment of the variability of groundwater levels and their causes;
- characterizing groundwater-level changes in time;
- assessment of the range of groundwater recharge;
- determining the reasons for groundwater level changes in Warsaw.

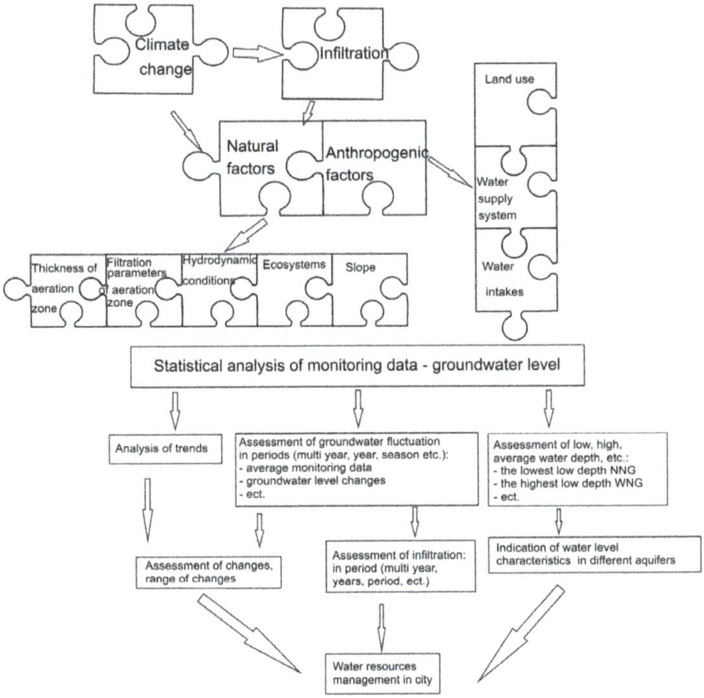

Figure 3. Factors and methods controlling groundwater infiltration and recharge in an urban area ([27], modified).

Measurements of the water table depth are made automatically with continuous control. The research station monitors three aquifers: two Quaternary aquifers (shallow and deeper) and the Oligocene aquifer (Figure 2), using 1 piezometer and 2 wells. The analysis was performed on a

27-year measurement series in hydrogeological years 1993–2019; therefore, the total number of daily observations was 9861 for each aquifer. In the frame of the groundwater monitoring system in Poland, the observation site of water levels is located in Warsaw (II-22-1, the western part of the city; Figure 1) in the deeper Quaternary aquifer. Measurements of groundwater levels have been performed at this site since 1994 [28].

Analysis of the changes of groundwater levels was performed using statistical methods most commonly applied in assessment studies of monitoring data [29–34]. The performed analysis allows the determination and characterization of statistical data and prediction of groundwater level changes.

Seasonal variability of the water level results from cyclic climate changes. It was characterized by annual amplitude, median value and analysis of charts with water table fluctuations. The annual amplitude was calculated as the difference between the highest and lowest multiannual mean water level, or as the difference between the absolute highest and the absolute lowest registered water level. The coefficient of variation, i.e., ratio of the standard deviation to the mean value, was also calculated. The measure of multiannual change of groundwater levels is the value of standard deviation of mean annual groundwater levels [35].

The magnitudes of estimated changes in the trend of meteorological variables in this study were estimated by the application of the Sen slope method. This technique calculates the gradient as a change in the measurements correlated with units of temporal change. The advantages of this method include an allowance for missing data, avoidance of assumptions about the distribution of tested data, and averting the effects of gross data errors and outliers [36]. Therefore, the method reduces the consequences of missing data and/or anomalous trends therein using the median values of the time series of various slopes that were detected as an evaluation tool [36,37]:

$$\beta = median \frac{(Xj - Xi)}{(j - i)} \text{ for all } j > i \tag{1}$$

In this equation, Xj and Xi denote values dated at times j and i, and time j is after time i. The estimator β is the median overall combination of the recorded pairs for the entire dataset where trend analysis is performed, and a positive β indicates an increasing trend, while a negative value indicates a decreasing trend [37].

Assessment of the trend of changes in the water table level in the studied aquifers using the Sen slope method was based on the average monthly positions of the water table.

The magnitude of infiltration was determined based on measurements of groundwater levels. The methodology of assessing the magnitude of recharge at a regional scale is the focus of many hydrogeological investigations, and their synthesis has been presented e.g., by de Vries, Simmers [38], Scanlon et al. [39], and Assouline [40].

Observations of the groundwater levels of the shallow Quaternary aquifer (first aquifer) allowed the determination of the recharge using the water table fluctuation (WTF) method (e.g., [41,42]). The method is based on an assumption that the increase in groundwater level is caused by infiltration. In intervals with no recharge, the water level decreases. Recharge is calculated as follows:

$$R = \frac{S_y \, dh}{dt} = \frac{S_y \, \Delta h}{\Delta t} \tag{2}$$

where

R—recharge [L/T] (mm/year)
S_y—specific yield [-]
h—hydraulic head [l] (mm)
t—time [T] (year)

The method is mainly used for short observation periods in areas with a shallow groundwater level, which causes quite large changes in its position over time [39].

The mean annual sum of precipitation from the multiannual interval 1993–2019 in the Warsaw area at 570 mm was used in the calculations of the infiltration coefficient [43]. The meteorological data came from the Warszawa-Okęcie Station, located in the south of Warsaw. A similar average amount of rainfall (571 mm) was recorded in the years 2001–2010 in Warsaw [44]

4. Results

The depth of the water table of the shallow Quaternary aquifer was at intervals from 3.62 to 7.76 m and a mean depth of 5.91 m over the period of 1993–2019. The deeper Quaternary aquifer had a depth of 6.40 to 10.13 m and a mean depth of 8.21 m. In the Oligocene aquifer, the extreme values of the depth to the water table were from 21.75 to 32.28 m, at a mean depth of 25.44 m (Table 1). The characteristic groundwater levels in three aquifers were determined in order to recognize the dynamics of changes indicating similarities and differences in the hydrodynamic conditions.

Table 1. Basic statistical parameters of the analyzed aquifers.

Parameters / Depth to Water Table *	Shallow Quaternary Aquifer—First Aquifer n = 9861	Deeper Quaternary Aquifer—Second Aquifer n = 9861	Oligocene Aquifer—Third Aquifer n = 9861
Average (m)	5.91	8.21	25.44
Median (m)	5.99	8.11	24.48
Maximum (m)	7.76	10.13	32.28
Minimum (m)	3.62	6.40	21.75
Amplitude (m) (difference between the absolute highest and the absolute lowest registered water level)	4.14	3.73	10.53
Amplitude (m) (difference between the highest and lowest multiannual mean water level)	1.38	2.22	4.62
1 quartile	5.56 (n = 2444)	7.78 (n = 2444)	22.51 (n = 2201)
3 quartile	6.28 (n = 2472)	8.70 (n = 2455)	27.46 (n = 2451)
Standard deviation (m)	0.71	0.64	3.22
Skewness (-)	−0.27	0.05	0.73
Coefficient of variation (%)	11.98	7.85	12.65

* based on daily measurements.

The value of the 1st quartile (5.56) and 3rd quartile (6.28) was exceeded over 2440 times in the case of depth to groundwater of shallow Quaternary aquifer (Table 1). Exceedance of the 1st quartile was found in two periods: March 1993 to October 1996 and June 2010 to October 2012. Exceedance of the 3rd quartile was found in two periods: September 2005–March 2007 and June 2016–October 2019. The value of the 1st quartile (7.78) and the 3rd quartile (8.70) was exceeded also over 2440 times in the case of depth of groundwater of the deeper Quaternary aquifer (Table 1). Exceedance of the 1st quartile occurred in two periods: March 2006 to October 2007 and June 2009 to August 2012. Exceedance of the 3rd quartile occurred in two periods: June 1996 to July 2000 and October 2017 to October 2019. The Oligocene aquifer is different compared to Quaternary aquifers in the case of the 1st and 3rd quartiles. The 1st quartile was exceeded from October 2013 to October 2019, while the 3rd quartile was exceeded from January 1993 to July 1999.

Taking into consideration the coefficient of variation, it can be seen that variability of depth of the groundwater table is statistically significant in the case of shallow Quaternary and Oligocene aquifers (Table 1).

A similarity can be observed between the Quaternary aquifers with regard to low groundwater depths. The amplitudes (WNG-NNG; highest low groundwater depth to lowermost low groundwater depth) were similar in the first and second aquifers and reached a maximum of 3 m (Table 2). In both aquifers, there were similar differences between the mean depth to the water table in the multiannual interval and the mean depth at low levels (SNG; average low groundwater depth). In the case of the Oligocene aquifer, the amplitude (WNG-NNG) exceeded 10.5 m, whereas the difference between the mean depth to the water table in the multiannual interval and at low levels was 1 m. This indicator points to the similarity of hydrodynamic conditions in the Quaternary aquifers and the distinctiveness of the Oligocene aquifer.

Table 2. Characteristic groundwater levels.

Characteristic Levels	Shallow Quaternary Aquifer—First Aquifer	Deeper Quaternary Aquifer—Second Aquifer	Oligocene Aquifer—Third Aquifer
Average groundwater depth in 1993–2019 (m) (daily data)	5.91	8.21	25.44
WNG—highest low groundwater depth (m b.g.s.) (annual data)	4.99	7.13	22.06
SNG—average low groundwater depth (m b.g.s.) (annual data)	6.28	8.66	26.44
NNG—lowermost low groundwater depth (m b.g.s.) (annual data)	7.76	10.13	32.28
Amplitude WNG-NNG (m)	2.77	3.00	10.53
Amplitude (m) (multiannual mean)	4.14	3.73	10.53
Amplitude (m) (annual mean)	0.69	0.73	1.38
Ratio of annual mean to multiannual mean	0.17	0.20	0.13

The ratio of annual mean to multiannual mean is 0.17 in the first aquifer, 0.20 in the second aquifer and 0.13 in the third aquifer, which indicates significant multiannual changes at small seasonal amplitudes.

A comparison of daily measurements from both Quaternary aquifers shows that larger amplitudes occur in the shallow aquifer (4.14 m) than in the deeper aquifer (3.73 m). Analysis of the water table position on an annual basis did not reveal cyclic changes related to seasonal fluctuations of the water table, with typical high levels in spring–summer and low levels in autumn–winter. The water table position was not affected by seasonal infiltration changes. There was no direct relationship between precipitation and groundwater levels of the Quaternary aquifers (Figure 4). The analysis of trends of long-term changes carried out using the Sen's slope method indicates a decreasing trend of the water table in a shallow Quaternary aquifer (Table 3).

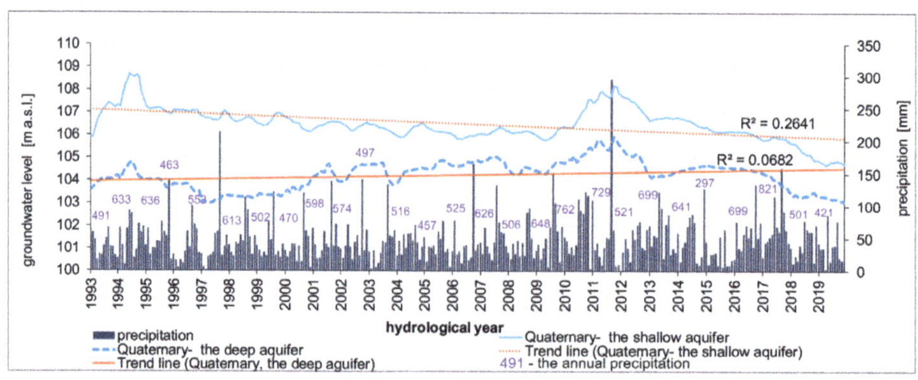

Figure 4. Monthly average groundwater levels of Quaternary aquifers.

Table 3. Sen's slope trend water table position.

Sen's Slope Trend	Shallow Quaternary Aquifer—First Aquifer	Deeper Quaternary Aquifer—Second Aquifer	Oligocene Aquifer—Third Aquifer	Deeper Quaternary Aquifer—Well No. II-22-1
Slope	−0.00433	0.00218	0.03045	0.00408
Lower	−0.00458	0.00192	0.02983	0.00384
Upper	−0.00411	0.00247	0.03102	0.00431

In the deeper Quaternary aquifer, the trend of changes of the water table position was opposite in the multiannual interval. However, detailed analysis of the groundwater levels in both aquifers shows that only in the first part of the multiannual interval (when the Quaternary aquifer was intensively exploited for industrial requirements), the water table of the deeper aquifer was at a lower level, indicating larger differences between both aquifers. After reducing the exploitation of groundwater due to the economic transformation in Poland in the early 1990s, since 1997 the water table of the deeper Quaternary aquifer began to increase systematically, and since 2001, its position has been analogous to the fluctuations in the shallow Quaternary aquifer. Similar dynamics of fluctuations in the multiannual interval 2001–2019 may indicate that after reducing exploitation in the deeper Quaternary aquifer, they are influenced by the same factors (recharge, drainage) that decide on the analogous water table position. The similarity between the water table positions of both Quaternary aquifers suggests the existence of a hydraulic connection between them in zones beyond the area of the research station (Figure 5).

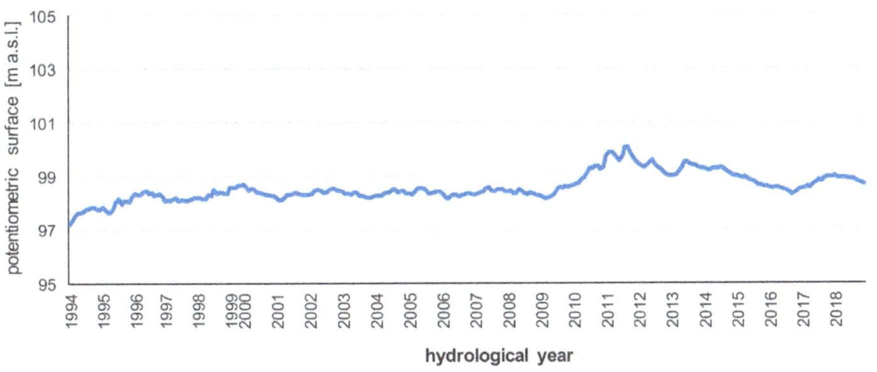

Figure 5. Monthly average potentiometric surface of the Quaternary aquifer in well no. II-22-1 [20].

Assessment of the reliability of accumulated data allowing for spatial interpretations in the Warsaw city was achieved by analysing changes of water table positions of the deeper Quaternary aquifer in well no. II-22-1 [28], located in the westernmost part of the city (Figure 5). Comparison of the water table fluctuations in both Quaternary wells indicates their clear similarity, which allows the local observations in the research station to be transferred to spatial assessments (Figures 4 and 5, Table 3).

The potentiometric surface of the Oligocene aquifer displays an opposite trend compared to both Quaternary aquifers. The trend of changes in the potentiometric surface of the Oligocene aquifer in the multiannual interval 1993–2019 points to the gradual restoration of hydrostatic pressure. In 1993, the potentiometric surface became stabilized at 80.83 m a.s.l., whereas it presently stabilizes at 90.72 m a.s.l., which shows a rise of the potentiometric surface in the multiannual interval by about 10 m (Figure 6).

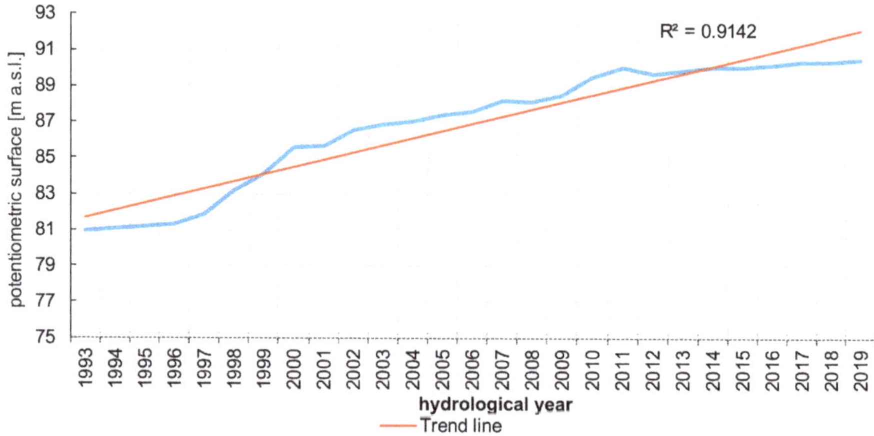

Figure 6. Monthly average potentiometric surface of Oligocene aquifer.

Trend analysis of the water table levels using Sen's slope method confirms the rising trend of the Oligocene potentiometric surface (Table 3).

The gradual restoration of pressure in the Oligocene aquifer results from reduced exploitation. The largest increase in the potentiometric surface was observed in 1997–2003 (Figures 6 and 7). After 2003, the potentiometric surface position changed in a much narrower range. There was excessive exploitation of groundwater from the Oligocene aquifer in Warsaw in the 1970s, reaching 50,000–60,000 m^3/d, resulting in the formation of a vast depression cone. Following the protection of these waters, in 1977, the construction of new water intakes for industrial purposes was banned, and the intake by existing industrial plants was strongly limited [45,46]. At present, 53 public sources rely on groundwater from Oligocene aquifer in the Warsaw city (as of 13.12.2019) [47]. Water from this aquifer is also used by hospitals and industries requiring water of the highest quality. The introduced restrictions have brought expected effects. A similar case took place in Barcelona, in which the intensive extraction of water from deltaic aquifers caused large drawdowns, leading to seawater intrusion [48]. Since the 1970s, many industries have migrated from the city, which reduced water demand, resulting in a progressive recovery of heads.

Accepting the urban area as the watershed-scale balance area and undertaking the attempt to identify the causes of groundwater level changes on a short-term and long-term basis, the basin water budget was evaluated for the shallow Quaternary aquifer. Recharge was assessed based on the analysis of changes of the water table position monitored in the first aquifer using the water table fluctuation (WTF) method. The specific yield taken for the calculations was at 0.10 [41,49–51]. This value is typical for sediments with low hydraulic conductivity (the aeration zone comprises sands with the addition of the silt fraction and clay).

Due to the lack of seasonal water table fluctuations, annual recharge was not discussed; instead, the method was modified by assessing recharge in intervals similar with regard to the dynamics of the water table. Based on the analysis of the water table position, four intervals were distinguished, differing in water table fluctuations as reflected in the magnitude of recharge (Figure 8).

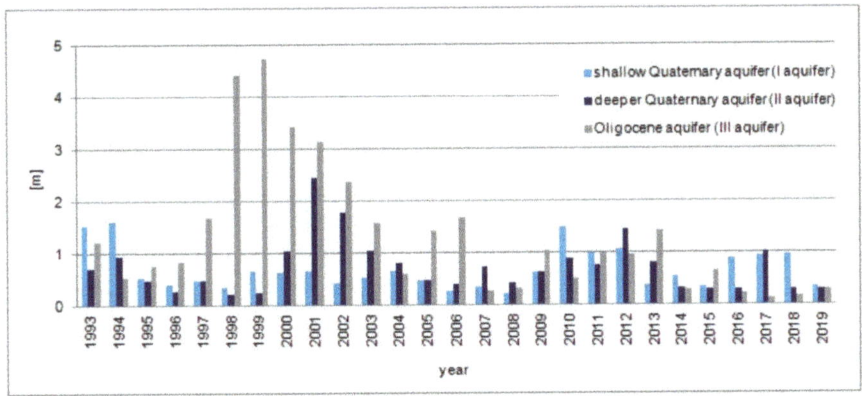

Figure 7. Annual amplitudes of the water table in Quaternary and Oligocene aquifers.

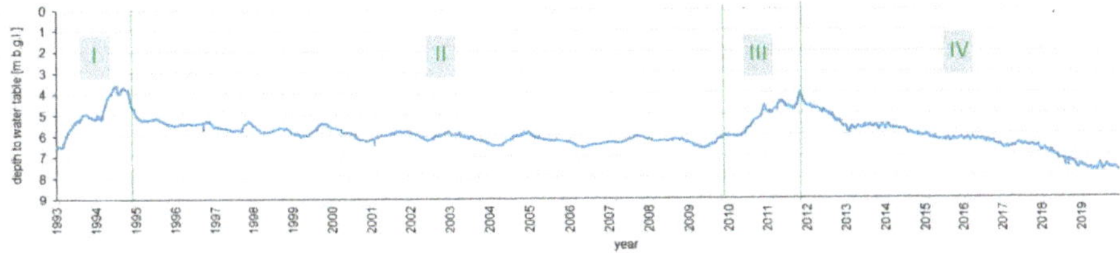

Figure 8. Water table in the shallow Quaternary aquifer in the selected research periods.

In the first interval, in the hydrological years 1993–1994, and in the third interval, in 2010–2011, infiltration was a long-term process (from 2 to 9 months), which caused a significant rise of the water table (1.2–2.9 m). The recharge coefficient was calculated as the percentage contribution of recharge (R) in the mean precipitation (P), assuming 570 mm/year of the precipitation for Warsaw. In these intervals, the annual recharge was from 130 to 160 mm, at a recharge coefficient from 23 to 28% (Table 4). In the second interval, in 1995–2009, the infiltration time was much shorter (1 to 2 months) and caused smaller seasonal fluctuations of the water table. Annual recharge was about 80 mm, and the recharge coefficient was 14%. For 16 years, the water table fell by almost 2 m. The fourth interval, in 2012–2019, was characterized by slightly lower values of parameters; infiltration was short-term (5–20 days), and despite a few such episodes during the year, systematic fall of the water table was observed. In 2019, the water table was at 7.71 m, i.e., 1.8 m below the mean value for the entire multiannual interval 1993–2019.

Table 4. Assessment of recharge of the shallow Quaternary aquifer using the WTF method.

Interval	Period	Depth to Water Table (m)		Infiltration Time (Days)		Infiltration during Water Table Rise (R) (mm)		Annual Recharge (mm)	Recharge Coefficient (%) (R/P)
		min	max	min	max	min	max		
I	1993–1994	3.62	6.51	133	254	155	160	~160	28
II	1995–2009	5.29	6.48	71	194	30	56	~80	14
III	2010–2011	4.88	6.08	61	286	84	120	~130	23
IV	2012–2019	6.39	5.86	5	20	5	17	~60	11

In the first aquifer, only long-term recharge lasting several months can result in changes in the water table position. The observed differences in the water table position and its gradual decrease from 2012 are not caused entirely by climate change as reflected in precipitation recharge. The distribution of annual totals of precipitation did not differ significantly from the remaining intervals during this period (Figure 4). This factor is not the only reason for long-term and non-cyclic changes. The lack of typical seasonality is the cumulative effect of many elements, which are interdependent and cannot be analyzed separately.

5. Discussion

Recharge in urban areas depends on a number of factors, including climate conditions and the dynamics of their changes [52,53], lithology of subsurface sediments [54,55], topography [56], land use and land cover [52,57], and thickness of the aeration zone [55,58].

The change of climate conditions in Warsaw was recorded as anomalously high precipitation in 2017 (821 mm) and anomalously low precipitation in 2015 (297 mm) (Figure 4). Measurements of groundwater levels in the research station have indicated that this change has no direct impact on the change of groundwater levels, even in the case of the shallowest Quaternary aquifer. In all analyzed aquifers, the influence of precipitation on water table fluctuations was insignificant due to the rather unfavourable filtration parameters in the aeration zone and the impact of urban infrastructure, especially in the subsurface zone. The reaction of the water table, even after intense precipitation, was greatly delayed. Multiannual trends of changes in the position of the water table in Quaternary aquifers and a stable trend of increasing potentiometric surface in the Oligocene aquifer are notable.

In urban areas, the magnitude of recharge depends on land use. The research station is located in Warsaw's centre and was developed at a time when there was hardly any infrastructure around it. However, in the 27-year interval, numerous changes in its surroundings have taken place, particularly related to the development of the University of Warsaw, Ochota Campus infrastructure, which may have had a significant impact on groundwater recharge and flow in the shallowest zone. The recharge zone also includes more distant regions, covered by typical urban housing. New multi-family buildings, parking lots and streets were constructed, modifying direct recharge from the surface, which is typical of urban agglomerations. The construction of multi-story buildings requiring deep foundations permanently modified the structure of groundwater flow [59,60]. Some buildings required drainage during construction, which periodically changed the hydrodynamic system. This factor is difficult to estimate because of construction activities in the urban area, but has a significant influence on the local position of the water table.

In urban areas with high water demands, the hydrodynamic conditions are modified due to significant, focused groundwater sources. There are only a few groundwater sources near the research station and in all of Warsaw; therefore, restoration of piezometric surfaces was documented in the Oligocene aquifer. Leakage of the water supply system and infiltration of water from the leaking network to the aquifer, documented even in the vicinity of the research station (Figure 9), can be observed in Warsaw.

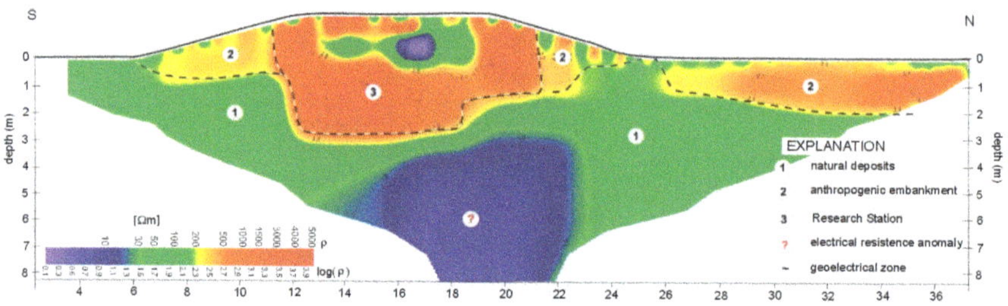

Figure 9. Electrical resistivity tomography profile in the research station area (January 2020).

The content of water leaking from the water supply system due to leakage or failure is roughly estimated in Warsaw at about 18% (mean value of water loss for Poland according to the Central Statistical Office for 2006), which gives the absolute value of 61,200 m^3 water per day. This value should be increased by another 18% resulting from leakage from the sewage system, which is 4160.7 km long in the Warsaw agglomeration [24]. Taking into account that these are estimates in many cases are understated, it can be concluded that a total of about 125,000 m^3 of water per day recharges the shallow Quaternary aquifer within this urban area. In Barcelona, the water supply network losses provide 22% of the total recharge of aquifers [48].

Water use does not increase proportionally with the increasing population of Warsaw. This may result from more economical water use due to rising costs and ecological lifestyles. This lower water source results in smaller amounts of water potentially recharging the subsurface zone due to leakage of the water system. The cumulative impact of many elements causing periodical rise or fall of recharge in urban areas allows for an assessment of their impact on the groundwater environment. In the case of the three analyzed aquifers, water level fluctuations are slow and occur over longer periods. Quaternary aquifers are characterized by multiannual intervals with higher recharge and much longer episodes of water table fall, which, compared to the entire 27-year interval, results in a falling trend towards decreasing water table levels. Such a case is very distinct in the shallow Quaternary aquifer since 2012. The Oligocene aquifer displays a trend of gradual but uneven restoration of hydrostatic pressure over the analyzed time span. The highest increase in the potentiometric surface was found in 1997–2003, after which the increase was much slower. The restoration of pressure in the Oligocene aquifer in the Warsaw area is confirmed by monitoring observations performed by the Polish Hydrogeological Survey [61].

6. Conclusions

Analysis of the changes in groundwater levels in several aquifers—two Quaternary ones and an Oligocene one—points to diverse factors controlling their location. In Quaternary aquifers, recharge conditions and dynamics of changes depend on local factors: infiltration conditions in the region of the monitoring system and development of the area, including elements characteristic of urbanized regions. In the case of the deep Oligocene aquifer, the position of the potentiometric surface depends on water circulation in the regional hydrogeological system, which covers a much larger area than the Warsaw agglomeration. Statistical analysis is a useful tool for interpreting multiannual measurements, although numerous natural and anthropogenic aspects need to be considered for its correct application. Groundwater monitoring systems usually exist in urban areas for several years, whereas statistical analysis requires long-term measurement series to analyse the trend of changes. The position of the water tables of the Quaternary aquifers across a long-term interval showed their fall, whereas the deep Oligocene aquifer displayed an opposite trend, with the restoration of hydrostatic pressure resulting in a rise in the potentiometric pressure by about 10 m.

Analysis of the changes in groundwater levels in time and factors influencing them in Warsaw does not show the impact of climate change (e.g., through extreme precipitation) on groundwater. Seasonal trends and long-term changes in levels have not been observed in the studied aquifers in the urban area.

The studies have shown that groundwater recharge and drainage in urban areas are components of natural processes related to hydrogeological and climate conditions and anthropogenic factors resulting from the groundwater source, water distribution, and changes in spatial management. When analysing the cumulative impact of these factors, often with opposite effects with regard to groundwater recharge, a relatively rapid decrease in the water table position may be observed in the shallowest groundwater aquifer. Therefore, it is critical to correctly manage water resources and increase water retention conditions through increased area covered by green zones, increased application of precipitation for urban green management and application of innovative solutions to enhance the infiltration of precipitation in high-density urban zones. Analysis of data sets containing long-term measurement series allows for a reliable assessment of the existing conditions (water efficiency), aids in planning sustainable water management and, to some degree, allows for the prediction of the effects of irrational water management or inappropriate development of urban infrastructure.

Author Contributions: Conceptualization, E.K. and J.J.M.; methodology, E.K., J.J.M., D.P. and A.W.; software, D.P. and A.W.; validation, E.K., J.J.M., D.P. and A.W.; resources, J.J.M. and A.W.; data curation, J.J.M. and A.W.; writing—original draft preparation E.K., J.J.M., D.P. and A.W.; visualization, D.P. and A.W. All authors have read and agreed to the published version of the manuscript.

Funding: This research received funding from the University of Warsaw.

Conflicts of Interest: The authors declare no conflict of interest.

References

1. Yang, L.; Feng, Q.; Yin, Z.; Wen, X.; Si, J.; Li, C.; Deo, R.C. Identifying separate impacts of climate and land use/cover change on hydrological processes in upper stream of Heihe River, Northwest China. *Hydrol. Proces.* **2017**, *31*, 1100–1112. [CrossRef]
2. Butler, D.; Digman, C.J.; Makropoulos, C.; Davies, J.W. *Urban Drainage*; Imprint CRC Press: Boca Raton, FL, USA, 2018; p. 592. ISBN 9781351174305. [CrossRef]
3. Fang, X.; Ren, L.; Li, Q.; Zhu, Q.; Shi, P.; Zhu, Y. Hydrologic response to land use and land cover changes within the context of catchment-scale spatial information. *J. Hydrol. Eng.* **2013**, *18*, 1539–1548. [CrossRef]
4. Gashaw, T.; Tulu, T.; Argaw, M.; Worqlul, A. Modeling the hydrological impacts of land use/land cover changes in the Andassa watershed, Blue Nile Basin, Ethiopia. *Sci. Total. Environ.* **2018**, *619–620*, 1394–1408. [CrossRef]
5. Ozler, H.M. Controlling seawater intrusion beneath coastal cities. In *Current Problems of Hydrogeology in Urban Areas, Urban Agglomerates and Industrial Centres*; Howard, K.W.F., Israfilov, R.G., Eds.; Nato Science Series (Series IV: Earth and Environmental Sciences); Springer: Dordrecht, The Netherlands, 2002; Volume 8. [CrossRef]
6. Foster, S. Global Policy Overview of Groundwater in Urban Development—A Tale of 10 Cities! *Water* **2020**, *12*, 456. [CrossRef]
7. Duque, J.; Chambel, A.; Madeira, M. The influence of urbanisation on groundwater recharge and discharge in the city of Evora, South Portugal. In *Current Problems of Hydrogeology in Urban Areas, Urban Agglomerates and Industrial Centres*; Nato Science Series (Series IV: Earth and Environmental, Sciences); Howard, K.W.F., Israfilov, R.G., Eds.; Springer: Dordrecht, The Netherlands, 2002; Volume 8. [CrossRef]
8. Tellam, H.; Thomas, A. Well water quality and pollutant source distributions in an urban aquifer. In *Current Problems of Hydrogeology in Urban Areas, Urban Agglomerates and Industrial Centres*; Nato Science Series (Series IV: Earth and Environmental, Sciences); Howard, K.W.F., Israfilov, R.G., Eds.; Springer: Dordrecht, The Netherlands, 2002; Volume 8. [CrossRef]
9. Magmedov, V.G.; Galetsky, L.S.; Yakovlev, Y.A. Urban groundwater contamination: Lessons from the Donbass Region, Ukraine. In *Current Problems of Hydrogeology in Urban Areas, Urban Agglomerates and Industrial Centres*; Nato Science Series (Series IV: Earth and Environmental Sciences); Howard, K.W.F., Israfilov, R.G., Eds.; Springer: Dordrecht, The Netherlands, 2002; Volume 8. [CrossRef]
10. Turkman, A.; Aslan, A.; Yilmaz, Z. Groundwater quality and pollution problems in-the Izmir Region of Turkey. In *Current Problems of Hydrogeology in Urban Areas, Urban Agglomerates and Industrial Centres*; Nato Science Series (Series IV: Earth and Environmental, Sciences); Howard, K.W.F., Israfilov, R.G., Eds.; Springer: Dordrecht, The Netherlands, 2002; Volume 8. [CrossRef]
11. Urban Europe—Statisticson cities, towns and suburbs—The dominance of capital cities. Available online: https://ec.europa.eu/eurostat/statistics-explained/pdfscache/50932.pdf (accessed on 10 May 2020).
12. Preciuk, WaterPIX/EEA. 2018. Available online: https://www.eea.europa.eu/pl/sygna142y/sygnaly-2018/artykuly/zuzycie-wody-w-europie-2014 (accessed on 10 May 2020).
13. WFD (Water Framework Directive). *Directive 2000/60/EC of the European Parliament and of the Council of 23 October 2000*; Official Journal of the European Communities L 327/1; WFD: Rome, Italy, 2000.
14. European Environment Agency. Water in the city, 2019 Copyright: D. Aydemir/EEA. Available online: https://www.eea.europa.eu/signals/signals-2018-content-list/articles/close-up-2014-water-in (accessed on 15 April 2020).
15. Lerner, D.N. Identifying and quantifying urban recharge: A review. *Hydrogeol. J.* **2002**, *10*, 143–152. [CrossRef]
16. Spalvins, A. Modelling as a Powerful Tool for Predicting Hydrogeological Change in Urban and Industrial Areas. In *Current Problems of Hydrogeology in Urban Areas, Urban Agglomerates and Industrial Centres*; Nato Science Series (Series IV: Earth and Environmental, Sciences); Howard, K.W.F., Israfilov, R.G., Eds.; Springer: Dordrecht, The Netherlands, 2002; Volume 8. [CrossRef]

17. Schirmer, M.; Leschik, S.; Musolff, A. Current research in urban hydrogeology: A review. *Adv. Water Resour.* **2013**, *51*, 280–291. [CrossRef]

18. Karpf, C.; Krebs, P. Modelling of groundwater infiltration into sewer systems. *Urban Water J.* **2013**, *10*, 221–229. [CrossRef]

19. Dohnalik, P. *Water Losses in Municipal Water Supply Networks*; PFOZW: Bydgoszcz, Poland, 2000. (In Polish)

20. Li, H.; Zhang, Y.; Vaze, J.; Wang, B. Separating effects of vegetation change and climate variability using hydrological modelling andsensitivity-based approaches. *J. Hydrol.* **2012**, *420*, 403–418. [CrossRef]

21. Małecki, J.J. The impact of urbanization on the specific levels of groundwater table–an analysis of the results of study conducted in the research station of the Faculty of Geology, University of Warsaw. *Biul. PGI* **2013**, *456*, 377–383. (In Polish)

22. Petrucci, G.; De Bondt, K.; Claeys, P. Toward better practices in infiltration regulations for urban stormwater management. *Urban Water J.* **2017**, *14*, 546–550. [CrossRef]

23. Dominko, L.; Kobyliński, A.; Kaliński, I.; Brodecki, A. *Hydrogeological Documentation of the Groundwater Basin of the Central Vistula Pradolina (GZWP–220)*; Arch. PGI-NRI: Warsaw, Poland, 1998. (In Polish)

24. Annual Report; Municipal Water and Sewage; Warsaw. 2017. Available online: https://mpwik.com.pl/download.php?id=2068 (accessed on 10 April 2020). (In Polish).

25. Office of Warsaw Infrastructure. The policy of water and sewage system development in the Capital City of Warsaw until 2025. In *Public Information Bulletin of the Capital City of Warsaw*; Ministry of Digital AffairsService Development Department: Warsaw, Poland, 2018. (In Polish)

26. Polish Geological Institute—National Research Institute. Available online: https://www.pgi.gov.pl/psh/zadania-psh/9010-zadanie-psh-zasoby-eksploatacyjne-wod-podziemnych.html (accessed on 10 April 2020).

27. Krogulec, E. Evaluating the risk of groundwater drought in groundwater-dependent ecosystems in the central part of the Vistula River Valley, Poland. *Ecohydrol. Hydrobiol.* **2018**, *18*, 82–91. [CrossRef]

28. PGI-NRI Database. Available online: http://spdpsh.pgi.gov.pl/PSHv7/ (accessed on 5 May 2020).

29. Helsel, D.R.; Hirsch, R.M. Statistical methods in water resources. In *Techniques of Water Resources*; Investigations of the United States; USGS: Reston, VA, USA, 2002; p. 510. [CrossRef]

30. Moon, S.K.; Woo, N.C.; Lee, K.S. Statistical Analysis of Hydrograph and Water-Table Fluctuation to Estimate Groundwater Recharge. *J. Hydrol.* **2004**, *292*, 198–209. [CrossRef]

31. Coppola, E.A.; Rana, A.; Poulton, M.; Szidarovszky, F. A neural network model for predicting water table elevations. *Ground Water* **2005**, *43*, 231–241. [CrossRef]

32. Shiria, J.; Kisi, O.; Yoon, H.; Lee, K.-K.; Nazemia, A.H. Predicting groundwater level fluctuations with meteorological effect implications-A comparative study among soft computing techniques. *Comput. Geosci.* **2013**, *56*, 32–44. [CrossRef]

33. Krogulec, E.; Zabłocki, S. Relationship between the environmental and hydrogeological elements characterizing groundwater-dependent ecosystems in central Poland. *Hydrogeol. J.* **2015**, *23*, 1587–1602. [CrossRef]

34. Krogulec, E.; Zabłocki, S.; Sawicka, K. Changes in groundwater regime during vegetation period in groundwater dependent ecosystems. *Acta Geol. Pol.* **2016**, *66*, 525–540. [CrossRef]

35. Chełmicki, W. Selected methods of assessing groundwater level fluctuations. *Pol. Geogr. Rev.* **1989**, *61*, 63–76. (In Polish)

36. Sen, P.K. Estimates of the regression coefficient based on Kendall's Tau. *J. Am. Stat. Assoc.* **1968**, *63*, 1379–1389. [CrossRef]

37. Deo, R.; McAlpine, C.; Syktus, J.; McGowan, H.; Phinn, S. On Australian heat waves: Time series analysis of extreme temperature events in Australia, 1950–2005. In Proceedings of the International Congress on Modelling and Simulation (MODSIM07), Christchurch, New Zealand, 10–13 December 2007; pp. 626–635.

38. De Vries, J.J.; Simmers, I. Groundwater recharge: An overview of processes and challenges. *Hydrogeol. J.* **2002**, *10*, 5–17. [CrossRef]

39. Scanlon, B.R.; Healy, R.W.; Cook, P.G. Choosing appropriate techniques for quantifying groundwater recharge. *Hydrogeol. J.* **2002**, *10*, 18–39. [CrossRef]

40. Assouline, S. Infiltration into soil: Conceptual approaches and solutions. *Water Resour. Res.* **2013**, *49*, 1–18.'[CrossRef]

41. Healy, R.W.; Cook, P.G. Using groundwater levels to estimate recharge. *Hydrogeol. J.* **2002**, *10*, 91–109. [CrossRef]

42. Krogulec, E. *Vulnerability Assessment of Groundwater Pollution in the River Valley on the Basis of Hydrodynamic Evidence*; Warsaw University Print: Warsaw, Poland, 2004. (In Polish)

43. Weather Online. Available online: https://www.weatheronline.pl/ (accessed on 10 January 2020).

44. Environment Statistical Analyses, Statistics Poland. 2019. Available online: https://stat.gov.pl/obszary-tematyczne/srodowisko-energia/srodowisko/ochrona-srodowiska-2019,1,20.html (accessed on 5 May 2020). (In Polish)

45. Macioszczyk, T.; Rodzoch, A.; Frączek, E. *Projektowanie stref ochronnych źródeł i ujęćwód podziemnych. (Designing Buffer Zones of Sources and Intakes of Ground Waters)*; Ministerstwo Ochrony Środowiska, Zasobów Naturalnych i Leśnictwa, Departament Geologii: Warsaw, Poland, 1993. (In Polish)

46. Nowicki, Z. (Ed.) *Groundwater of Voivodship Cities in Poland*; PGI: Warsaw, Poland, 2007; ISBN 978-83-7538-152-8. (In Polish)

47. City of Warsaw. Available online: https://warszawa19115.pl/-/ujecia-wody-oligocenskiej (accessed on 5 March 2020).

48. Vazquez-Sune, E.; Carrera, J.; Tubau, I.; Sanchez-Vila, X.; Soler, A. An approach to identify urban groundwater recharge. *Hydrol. Earth Syst. Sci.* **2010**, *14*, 2085–2097. [CrossRef]

49. Crosbie, R.S.; Binning, P.; Kalma, J.D. A time series approach to inferring groundwater recharge using the water table fluctuation method. *Water Resour. Res.* **2005**, *41*. [CrossRef]

50. Gehman, C.l.; Harry, D.L.; Sanford, W.E.; Stednick, J.D.; Beckman, N.A. Estimating specific yield and storage change in an unconfined aquifer using temporal gravity surveys. *Water Resour. Res.* **2009**, *45*, W00D21. [CrossRef]

51. Walker, D.; Parkin, G.; Schmitter, P.; Gowing, J.; Tilahun, S.A.; Haile, A.T.; Yimam, A.Y. Insights from a Multi-Method Recharge Estimation Comparison Study. *Groundwater* **2019**, *57*, 245–258. [CrossRef]

52. Crosbie, R.S.; Scanlon, B.R.; Mpelasoka, F.S.; Reedy, R.C.; Gates, J.B.; Zhang, L. Potential climate change effects on groundwater recharge in the High Plains Aquifer, USA. *Water Resour. Res.* **2013**, *49*, 3936–3951. [CrossRef]

53. Meixner, T.; Manning, A.H.; Stonestrom, D.A.; Allen, D.M.; Ajami, H.; Blasch, K.W.; Brookfield, A.E. Implications of projected climate change for groundwater recharge in the western United States. *J. Hydrol.* **2016**, *534*, 124–138. [CrossRef]

54. Acharya, S.; Jawitz, J.W.; Mylavarap, R.S. Analytical expressions for drainable and fillable porosity of phreatic aquifers under vertical fluxes from evapotranspiration and recharge. *Water Resour. Res.* **2012**, *48*, W11526. [CrossRef]

55. Cao, G.; Scanlon, B.R.; Han, D. Impacts of thickening unsaturated zone on groundwater recharge in the North China Plain. *J. Hydrol.* **2016**, *537*, 260–270. [CrossRef]

56. Huang, H.; Huang, T.; Pang, Z.; Liu, J.; Yin, J.L.; Edmunds, W.M. Groundwater recharge in an arid grassland as indicated by soil chloride profile and multiple tracers. *Hydrogeol. Process.* **2017**, *31*, 1047–1057. [CrossRef]

57. Scanlon, B.R.; Reedy, R.C.; Gates, J.B. Effects of irrigated agroecosystems: 1. Quantity of soil water and groundwater in the southern High Plains, Texas. *Water Resour. Res.* **2010**, *46*. [CrossRef]

58. Zhang, G.H.; Fei, Y.H.; Yang, L.Z. Influence of unsaturated zone thickness on precipitation infiltration for recharge of groundwater. *J. Hydraul. Eng.* **2007**. Available online: http://en.cnki.com.cn/Article_en/CJFDTotal-SLXB200705015.htm (accessed on 5 May 2020).

59. Attard, G.; Winiarski, T.; Rossier, Y.; Eisenlohr, L. Review: Impact of underground structures on the flow of urban groundwater. *Hydrogeol. J.* **2015**, *24*, 5–19. [CrossRef]

60. Opęchowski, K.; Krogulec, E. Impact of deep foundations on the groundwater flow structure in the hydrogeodynamically vulnerable area. *Geol. Rev.* **2019**, *67*, 478–486. (In Polish) [CrossRef]

61. Karta Informacyjna JCWPd 65. Available online: https://www.pgi.gov.pl/dokumenty-pig-pib-all/psh/zadania-psh/jcwpd/jcwpd-60-79/4425-karta-informacyjna-jcwpd-nr-65/file.html (accessed on 5 May 2020). (In Polish)

Publisher's Note: MDPI stays neutral with regard to jurisdictional claims in published maps and institutional affiliations.

Article

Modelling the Role of SuDS Management Trains in Minimising Flood Risk, Using MicroDrainage

Craig Lashford [1],*, Susanne Charlesworth [1], Frank Warwick [2] and Matthew Blackett [1]

[1] Centre for Agroecology, Water & Resilience, Coventry University, Coventry CV1 5FB, UK; apx119@coventry.ac.uk (S.C.); aa8533@coventry.ac.uk (M.B.)

[2] School of Energy, Construction & Environment, Coventry University, Coventry CV1 5FB, UK; aa4510@coventry.ac.uk

* Correspondence: ab0874@coventry.ac.uk

Received: 6 August 2020; Accepted: 4 September 2020; Published: 13 September 2020

Abstract: This novel research models the impact that commonly used sustainable drainage systems (SuDS) have on runoff, and compare this to their land take. As land take is consistently cited as a key barrier to the wider implementation of SuDS, it is essential to understand the possible runoff reduction in relation to the area they take up. SuDS management trains consisting of different combinations of detention basins, green roofs, porous pavement and swales were designed in MicroDrainage. In this study, this is modelled against the 1% Annual Exceedance Potential storm (over 30, 60, 90, 120, 360 and 720 min, under different infiltration scenarios), to determine the possible runoff reduction of each device. Detention basins were consistently the most effective regarding maximum runoff reduction for the land they take (0.419 L/s/m^2), with porous pavement the second most effective, achieving 0.145 L/s/m^2. As both green roofs (20.34%) and porous pavement (6.76%) account for land that would traditionally be impermeable, there is no net-loss of land compared to a traditional drainage approach. Consequently, although the modelled SuDS management train accounts for 34.86% of the total site, just 7.76% of the land is lost to SuDS, whilst managing flooding for all modelled rainfall and infiltration scenarios.

Keywords: detention basins; green roofs; MicroDrainage; porous pavement; runoff reduction; swales

---

1. Introduction

1.1. Sustainable Drainage Systems Management Train

In an evolving environment, the risk of flooding is increased when conventional piped based systems are not adapted to manage the intensified stormwater runoff associated with both increased impermeable surfaces and a changing climate. Sustainable drainage systems (SuDS), however, provide an alternative approach to managing stormwater and flooding [1]. SuDS mimic natural hydrological processes which have been lost, due to urbanization and associated impermeable surfaces and the installation of pipe-based drainage [2]. Whilst the SuDS square emphasizes the equal importance of water quantity, water quality, amenity and biodiversity, regarding the likely benefits of integrating SuDS, in reality, water quantity or quality benefits are typically prioritized by drainage engineers [3]. A SuDS management or treatment train is a system which utilizes a range of SuDS devices in sequence to reduce flow and the overall level of pollution in runoff [4]. When focusing on water quality, combinations of SuDS devices are often termed treatment trains, however, the focus of the present study is on water quantity reduction, and therefore, the term "SuDS management train" will be used.

A SuDS management train provides extra levels of resilience against flooding, as more devices are used, resulting in greater levels of water retention [5]. It is not always feasible to utilize one large device at a site, therefore, a series of smaller linked devices in a management train can be more practical, meeting the requirements of the SuDS square. Past research has demonstrated the effectiveness with which a SuDS management train can reduce runoff, in comparison with conventional drainage, and in comparison to impermeable surfaces [6]. Limited research exists regarding which devices should be prioritized in a management train and how different devices work when combined, although susdrain [7] suggested that swales, detention basins, green roofs and porous pavement are amongst the most commonly used in England and Wales.

1.2. Sustainable Drainage Management Trains: Barriers to Their Implementation

Although research demonstrates the benefits of integrating SuDS, there has been a reluctance from stakeholders and practitioners in the UK to see SuDS implemented as an approach to more sustainable methods of drainage [8]. SuDS are a divergence from traditional pipe-based methods, encouraging the integration of more natural open water management in the built environment, and such a change in paradigm has resulted in resistance to their wider implementation [2].

The primary concern of developers typically relates to the design, integration and operational role of SuDS, with a perceived lack of clarity on the potential benefits that SuDS can bring [9,10]. Land use is a key barrier, particularly at new build sites where there is often an emphasis on maximizing profits by housebuilders and construction companies, and therefore, minimizing above-ground drainage space [8,11,12]. Maximizing space is equally critical when redesigning urban areas and as such, ensuring the most effective SuDS devices are identified is vital [13]. The innovative research presented here incorporates a management train approach in demonstrating how effective different SuDS approaches can be, in terms of their impact on runoff and their footprint, and how beneficial SuDS can be for flood risk management. UK flooding in winter 2015/2016 impacted 16,000 properties and caused GBP 1.3 billion damage; given the UK Government's commitment to building 300,000 new homes a year until 2023, SuDS offer a sustainable approach to future flood management, particularly in a changing climate [14]. Climate change is expected to increase annual rates of rainfall, as well as both intensity and frequency of large rainfall events, particularly during winter [15]. Consequently, the annual cost of flooding in the UK is expected to rise under the 2deg warming scenario by 54% by 2080, with 40% more properties likely to be impacted by flooding by 2080, even with the assumption that population remains static [16].

To demonstrate the benefits, and promote the wider integration of SuDS into UK drainage design, this research focuses on the water quantity benefits of different SuDS techniques, and their benefits in relation to land-take, particularly when combined in a management train. The project compares the benefits of SuDS installations to the associated loss of land, focusing on two key areas of resistance: Peak runoff reduction and land take. Whilst much of the research focuses on new-build developments, the prioritization of space and the impact by area and volume of different devices is pertinent for retrofit installations; previous studies have not standardized the impact on runoff of different linked devices by area (m^2) and volume (m^3) [17]. This research provides an assessment of the most commonly used SuDS in the UK and how effectively they reduce runoff in relation to the amount of land take.

2. Methods

Analysis of Sustainable Drainage Systems in a Management Train

This study uses the UK industry standard drainage model, MicroDrainage v2019.1 [18], and adapts the method from Lashford et al. [6], which determined the overall modelled runoff reduction of using SuDS management trains. The research framework for this study is presented in Figure 1. This research further assesses each device regarding the space they account for in the model. A SuDS management train was developed using the DrawNet tool, with LiDAR data with a 1 m^2 vertical resolution used to determine flood flow routes (Figure 1). Therefore, a site was designed based nominally on an existing 5 ha new build development with 250 proposed houses, in Coventry, UK (Figure 2). The SuDS devices used in this study are presented in Table 2, and constitute the most commonly used devices in management trains in England and Wales, based on an analysis of the susdrain [7] database, with the designed layout based on existing SuDS management trains in Hamilton, Leicester, UK and Upton, Northampton, UK [3]. The area and volume presented in Table 2 were calculated based on the overall site design. Although the focus of the findings is on the total area (m^2), as space is defined as a key barrier to the uptake of SuDS, to understand the characteristics of the SuDS used, the total storage volume (m^3) is also provided (Figure 1) [8–10].

Table 1. Different combinations of each SuDS device in the modelled management trains.

Devices Used
Swale
Green roof and swale
Porous pavement and swale
Green roof, porous pavement and swaleSwale and detention basin
Green roof, swale and detention basin
Porous pavement, swale and detention basin
Green roof, porous pavement, swale and detention basin

Table 2. Total area, % of the total land take and volume of each SuDS device used in the analysis.

Device	Total Area (m^2)	% of Total Land Take	Total Volume (m^3)
Detention basin	2189	4.38	4658
Green roof	10,170	20.34	1017
Porous Pavement	3380	6.76	1568
Swale	1692	3.38	1322

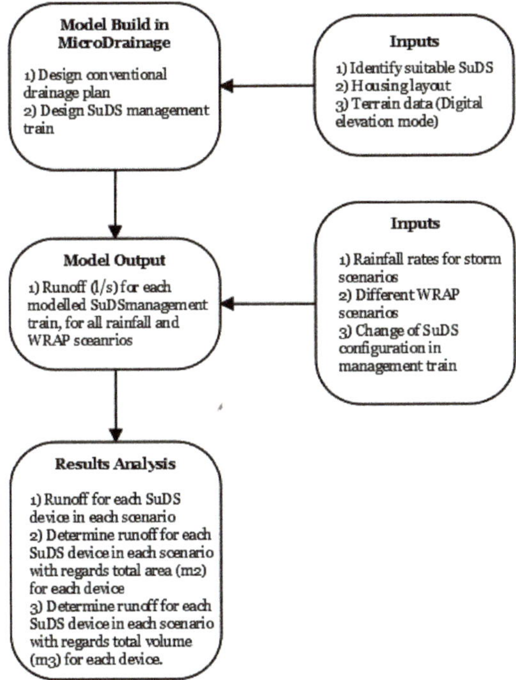

Figure 1. The research framework for the project.

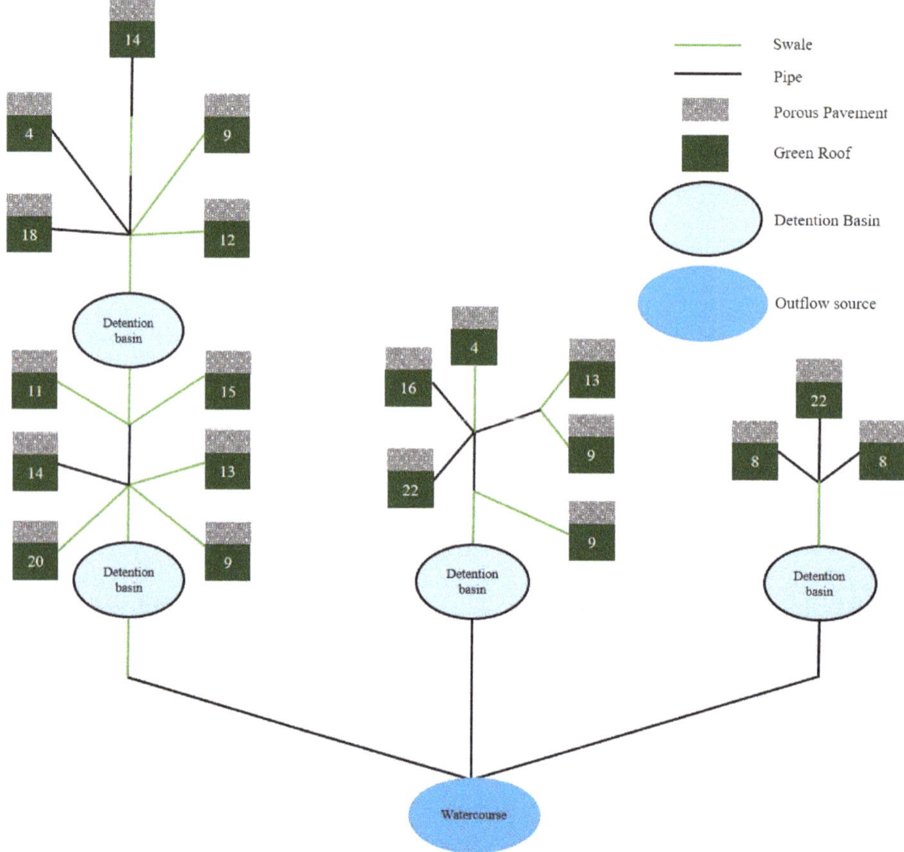

Figure 2. Schematic layout of the SuDS management train, not to scale. If a SuDS device was not included in the model (Table 1), they were bypassed. The number indicates the total number of houses at the site.

The modelled porous pavement, which were integrated on the driveway for each house, had an infiltration coefficient consistent with overall site infiltration, and therefore, varied dependent on the winter rainfall acceptance potential (WRAP) value. MicroDrainage uses the WRAP method to quantify infiltration and standard percentage runoff and was developed in the Flood Studies Report categorizing soil based on soil water regime, depth to an impermeable layer, the permeability of soil horizons and slope of land [19–21]. To ensure the wider applicability of the findings, the WRAP of 0.5 (low infiltration) 0.3 (medium infiltration) 0.15 (high infiltration) were used. The porosity of each device was 0.3, as suggested by Woods Ballard et al. [3], with a maximum membrane percolation of 1000 mm h^{-1} and a total depth of 450 mm, in line with British Standard Institution [22] specifications. The green roofs were designed on each of the 250 houses, with 5 mm depression storage, an overall depth of 100 mm and a decay coefficiency 0.05, as recommended by Stovin [23]. The governing equation for flow through the green roof is based on the assumption made by XP Solutions that the roof will become saturated at 120 min, regardless of rainfall intensity. Swales were designed with a Manning's value of 0.06, in accordance with Chow [24], and were designed to run alongside the proposed pavement structure, ranging from a width of 600 mm to 3000 mm, when space was available (Figure 2). Wider swales were used in this study away from the roadside, when conveying flow from detention basins. Designing swales alongside pavements reduces the amount of open space that the SuDS devices occupy [17]. For safety reasons swales, had a 1:3 side slope, which limited gradient [3].

$$A_t = e^{-kt}, \; a_t = \left(\frac{A_t}{\sum A_t}\right)a \tag{1}$$

where a is the total area of the green roof, A is a factor required to scale the curve to provide the correct total catchment area, e is the exponential, k is the decay coefficiency and t is time in min.

All source control devices were channelled into detention basins via swales or pipes. Pipes were used whenever open-channel flow was not possible, for example, when water was conveyed below a road (Figure 2). Once runoff from a source control device was collected in a detention basin, it was conveyed to a nearby watercourse. Swales were used in all scenarios to convey runoff. The impact of each device was compared with the total land take (m^2) and storage volume of each SuDS device (m^3) to determine the most space-effective design for the maximum and minimum achieved reduction of peak flow, dependent on the combined devices.

The different modelled management trains are highlighted in Table 1. Each combination was modelled for the 0.5, 0.3 and 0.15 WRAP values, with the overall site assigned a consistent WRAP value, apart from the impermeable surfaces, such as roads and pavements. The model was also run under differing rainfall durations, to understand the response to a range of rainfall intensities. The UK Non Statutory-Standards for SuDS require sites to be designed to deal with runoff for the 1% Annual Exceedance Potential (AEP) 360 min (11.92 mm/h) storm, therefore, this rainfall scenario was used in this research [25]. However, to demonstrate the effectiveness of devices to a variety of storm intensities, the 1% AEP 30 min (73.14 mm/h), 60 min (44.09 mm/h), 90 min (32.79 mm/h), 120 min (26.58 mm/h) and 720 min (7.18 mm/h) duration storms were also simulated.

3. Results

Figures 3–5 present the impact for each of the four modelled devices, dependent on the WRAP infiltration scenario and rainfall duration, based on the range of possible peak flow reduction (Figure 3a–d), the impact the device has in relation to area (L/s/m^2) (Figure 4a–d) and volume (L/s/m^3) (Figure 5a–d). Based on this modelling, maximum runoff reduction in comparison with conventional piped drainage was achieved for all SuDS components when used just in conjunction with swales (and for swales alone), for example, detention basins and swales. Figure 3a shows that detention basins, as part of the management train, consistently reduced maximum peak runoff (L/s) under all infiltration and rainfall scenarios more than any other modelled SuDS technique. Figure 3a–d show that as rainfall intensity decreases, the total reduction in runoff decreases. Furthermore, for all scenarios, the maximum reduction is achieved during the WRAP 0.5 event.

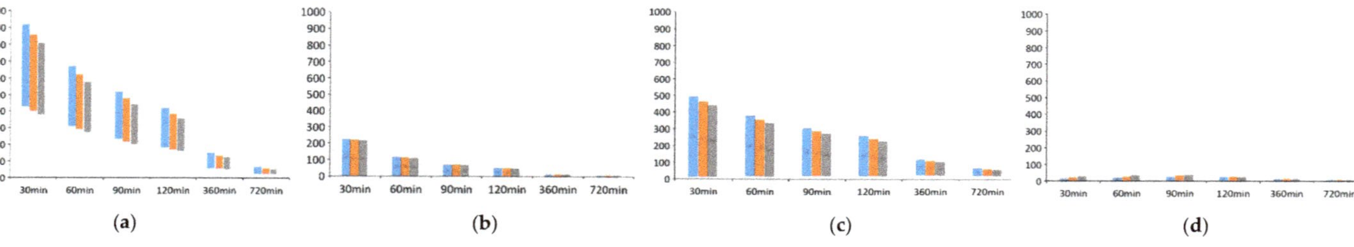

Figure 3. The maximum and minimum range of possible reduction of runoff (L/s) for each device; (a) Detention basins, (b) green roofs, (c) porous pavement (d) swales. Each color represents a different WRAP scenario. Blue 0.5, Orange 0.3 and Grey 0.15.

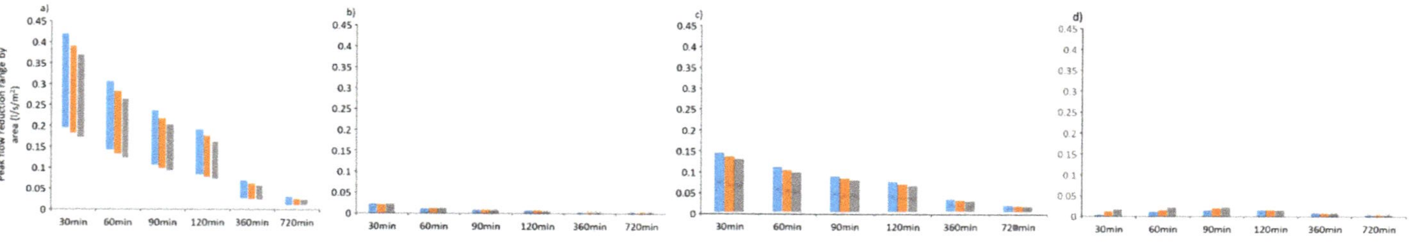

Figure 4. The maximum and minimum range of peak flow reduction by area (L/s/m^2) for each device; (a) Detention basins, (b) green roofs, (c) porous pavement (d) swales. Each color represents a different WRAP scenario. Blue 0.5, Orange 0.3 and Grey 0.15.

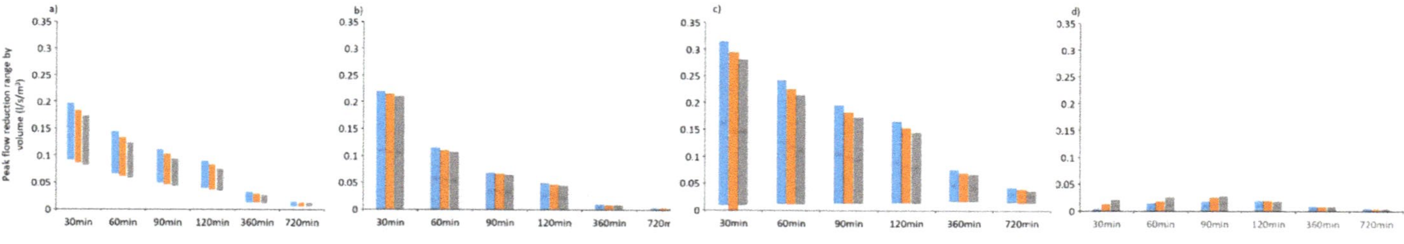

Figure 5. The maximum and minimum range of peak flow reduction by volume (L/s/m^3) for each device; (a) detention basins (b) green roofs (c) porous pavement (d) swales. Each color represents a different WRAP scenario. Blue 0.5, Orange 0.3 and Grey 0.15.

Detention basins were the most effective device for all scenarios when analyzing the impact on the area, achieving a maximum reduction of 0.419 L/s/m^2 for the 30 min storm under 0.5 WRAP infiltration conditions (Figure 4a). Conversely, Figure 5a shows that detention basins are less effective at reducing runoff, compared to the other modelled devices, in relation to their volume (L/s/m^3). This is likely due to their greater capacity to store water and increased total volume (Table 2).

The second consistently most effective device was porous pavement, which achieved a reduction of 0.145 L/s/m^2 for 30 min, 0.5 WRAP scenario (Figure 4c). However, when accounting for the total volume of the device (L/s/m^3), porous pavement outperformed the other modelled devices for all scenarios (Figure 5a–d). Although the modelled porous pavement covered 1191 m^2 (2.54% more) more total land than detention basins, they had a significantly reduced total volume (3090 m^3 less) than the modelled detention basins, accounting for the switch in performance depending on whether area or volume was analyzed (Table 2).

The additional source management device modelled was the green roof, which had little effect on reducing peak flow in comparison to the porous pavement for all rainfall durations greater than 120 min. The greatest peak flow reduction that was achieved for rainfall duration longer than 120 min, when adding green roofs to the swale management train, was 9 L/s, for the 360 min 0.5 WRAP model (Figure 3b), resulting in consistently less than 0.001 L/s/m^2 (Figure 4b) in relation to the area covered. However, the equation used in MicroDrainage that dictates flow in a green roof is based on a saturation capacity being reached at 120 min (see Equation (1)). The influence of the equation is evident in the runoff reduction for events up to and including 120 min (Figure 3b), where the performance of green roofs is greatly improved in comparison to the longer duration events. Consequently, their impact on reducing runoff decreased as storm intensity decreased, due to their significantly reduced storage capacity in comparison to other modelled SuDS techniques (at least 305 m^3 less total volume). Therefore, for each rainfall scenario up to and including the 120 min duration, green roofs outperformed swales, and for the intense 30 min storm, outperformed detention basins regarding peak reduction by total volume (L/s/m^3) (Figure 5). Swales were marginally more effective than green roofs, based on area utilized in the model design, for all storm durations over 60 min (Figure 4). Nevertheless, the influence of swales to reduce runoff remained limited for all scenarios; with a peak runoff reduction of 35.2 L/s occurring for the 90 min duration storm, 0.15 WRAP model, compared to 65 L/s runoff reduction for the same model configuration for green roofs (Figure 3).

Figures 3–5 show that when comparing infiltration scenarios and rainfall intensities, there is little modelled change for the role of the most effective devices; detention basins were consistently the most effective in terms of area and porous pavement in terms of volume, regardless of the 120 min delay in saturation for green roofs. The maximum reduction in peak flow for each device decreased as infiltration increased, apart from swales (Figures 3d, 4d and 5d). For detention basins, green roofs and porous paving, this is likely a result of their reliance on storing stormwater and less runoff entering all SuDS devices, with a greater amount being infiltrated into the surrounding green space, reducing the impact of each SuDS technique. However, swales act differently; runoff reduction (L/s) increases as infiltration increases, up to the 90 min storm duration, and plateaus for the 120 min storm onwards, with no discernible reduction regardless of infiltration rates.

4. Discussion

The findings presented in Figures 3–5 show that SuDS can substantially reduce total site runoff, and ultimately reduce flooding. As identified previously, space is a critical factor in determining the wider implementation of SuDS, detention basins and porous pavement should, therefore, be prioritized when implementing SuDS as part of a management train [8–10]. This provides drainage engineers with guidance when developing new sites, but also regarding urban retrofit flood risk management, where space is a premium; which has also been identified as a barrier to the wider implementation of SuDS [11,12]. Both devices were consistently the most effective at reducing runoff in relation to their land-take, accounting for 4.38% and 6.78% (Table 2) of the total modelled site, respectively

(Figures 3–5). Both detention basins and porous pavement encourage infiltration by storing large volumes of runoff during high-flow events, however, their effectiveness is largely constrained by site infiltration conditions, structure size, and in the case of porous pavement, void space and material used (as also document in Yazdi [26], and Scholz and Grabowiecki [27]).

Whilst detention basins were most effective in terms of their area, they account for 2189 m^2 of space that could otherwise be used for housing, decreasing potential site profits; a typical barrier to their integration in drainage schemes [10]. Furthermore, 83% of the population in England currently live in existing urban areas. With a likely rise of high-intensity rainfall events as a result of climate change, sustainable approaches to urban pluvial flood management are required [28]. Subsequently, integrating detention basins in an already densely populated space, where land is at a premium, is a challenge [29,30]. Figure 3a–d, however, demonstrate the reduction in runoff that can be achieved through including detention basins into urban design, and if designed effectively, can be successfully incorporated into both new build and retrofit sites. The UK has taken a risk-averse approach to open stormwater basins in cities, with few retrofit examples, preferring to install shallow, densely vegetated detention basins predominantly at new build sites, which are less multi-functional [9,31–33]. Nevertheless, there are successful international examples of integrating SuDS into urban design, for example, Malmo, Sweden and Portland, USA, where detention basins not only provide water quantity improvements, but retain multiple benefits, such as recreational space, when dry [34,35]. Although detention basins are not the entire solution to reducing overall flood risk, they present an opportunity for reducing local, pluvial flooding by storing large volumes of water [36]. Integrating above-ground visible strategies for flood alleviation, by developing SuDS management trains at site scale can also enhance community resilience to flooding, by increasing local understanding of flooding, and removing the traditional 'out of sight out of mind' mentality [37].

Porous pavement is used instead of traditional impermeable surfaces, and is often indistinguishable to the lay public [27]. In the model used for this research, porous pavement did not take up potential green space as it replaced existing driveways for each property. For this reason, regardless of the 6.76% total land take, there was no net-loss of land use, and there was a reduction of runoff when used in combination with other devices of up to 491.7 L/s, for the 30 min rainfall event (Figure 3c). Consequently, porous pavement is a suitable sustainable method of flood risk reduction, as it not only can reduce runoff, but also provides multiple uses, such as car parks and low traffic roads, increasing its amenity benefits [3]. However, the impact of porous pavement on runoff, when analyzed in terms of area, was consistently less than detention basins. This is a result of a greater modelled surface area for porous pavement, in comparison to detention basins, although the total below-ground storage capacity for detention basins was far greater (see Table 2). As detention basins were much deeper (1500 mm in comparison to a storage capacity of porous pavement of 450 mm) than other devices modelled, the area occupied was relatively modest. Therefore, they were nearly twice as effective for an equivalent space as porous pavement. This is further demonstrated in the analysis regarding volume, and explains why porous pavement, which had nearly three times less total volume than detention basins, were consistently more effective at reducing peak flow by total volume. Analysis of both area and volume shows the need to prioritize porous pavement at the source control stage and detention basins as site control, in addressing runoff, supporting research by Chen et al. [38] and Woods Ballard et al. [3] who endorsed both as highly effective flood management tools. Whilst the modelled detention basins take up 6.76% of modelled land, which could be used for additional housing and is typically presented as a barrier to the wider implementation of SuDS, the possible benefit on reducing runoff is considerable [8–10]. Similar to detention basins, there has been a risk-averse approach to integrating porous pavement in cities in the UK, due to a lack of experience of their design and implementation, the associated costs of retrofitting porous pavement and the possible disruption pre and post-installation [33]. However, whilst the concerns exist, particularly surrounding the cost of porous pavement, research by Gordon-Walker et al. [39] shows that there are cost benefits for using the device for runoff management. Furthermore, this research has demonstrated the scale of runoff reduction that can be achieved by utilizing porous pavement in the SuDS management train as an alternative to impermeable surfaces.

The other modelled source control device were green roofs, which has a relatively limited impact on reducing runoff in MicroDrainage, particularly as the modelled storm duration increases (Figure 3b). As green roofs have only a small storage depth of 100 mm, as suggested by Stovin [23], compared to the larger storage capabilities of detention basins and porous pavement, the effectiveness of green roofs is lower. However, the governing Equation (1) for flow in a green roof in MicroDrainage is based on the assumption that it reaches saturation at 120 min, regardless of total rainfall depth. Such an arbitrary storage capacity is an unrealistic representation of a green roof structure. Figure 3b shows that the green roof model reduces peak flow most effectively for the highest intensity 30 min storm (73.14 mm/h), with a reducing impact as intensity decreases. This is a result of Equation (1) in the model, which results in near total water storage in a green roof, regardless of the rainfall intensity. Using physical models, Stovin et al. [40], demonstrated that green roofs act entirely differently to the MicroDrainage model, with peak performance and runoff reduction achieved during low-intensity events, their abilities reducing as intensity increases. As green roofs are often sloped, their ability to retain water, particularly during high-intensity rainfall is restricted [23]. Whilst specific vegetation on a roof, if extensive, can slow down runoff, increasing infiltration, the influence of slope can considerably reduce the effectiveness of a green roof [41]. Nonetheless, all models show that green roofs are capable of reducing runoff, with varying levels of success, which is similar to the findings of De Ville et al. [42]. Therefore, while green roofs have limited potential in this example on reducing runoff, they still provide extra resilience to further increase the impact of a SuDS management train, according to MicroDrainage, in small events. Green roofs also have multiple benefits by greatly increasing the amenity and biodiversity potential of a site and are capable of cooling urban areas, reducing the impact of the urban heat island effect and providing carbon sequestration [43,44]. For these reasons, site requirements must be assessed to better understand the suitability of green roofs in site design. Green roofs can be considered at new build sites during the early phases of design, with buildings appropriately designed to accommodate the increased potential load.

Swales were capable of reducing runoff by up to 2.6 L/s more than green roofs for all infiltration scenarios, for storms with a longer duration than 120 min. However, as identified in Section 3, the performance of the modelled swales was in contrast to the other devices; their effectiveness increased as infiltration rates increased. Whilst the other modelled devices rely on storing water, with varying levels of success, swales are primarily used for conveying stormwater around a site [3]. The Manning's value used for swales was 0.06, therefore, whilst flow speeds are likely to be reduced, with some level of detention and consequently infiltration occurring, particularly in comparison to traditional impermeable pipes, the total runoff reduction is negligible [45]. Winston et al. [46] suggest that swales effectiveness for moderate scale events can be greatly improved by integrating check dams to further limit runoff and increase storage of stormwater. Additionally, their effectiveness is limited by the infiltration rate of a site, which corresponds with the modelled results for the higher intensity events [47,48]. The reduction in runoff for the two lowest intensity rainfall events (360 and 720 min) was negligible, ranging from a maximum of 1.3 L/s to 0.1 L/s. Therefore, the lack of storage in the modelled swales reduced their effectiveness, with limited runoff leaving the system through infiltration, in comparison with storing large volumes of water and allowing for slow infiltration, as occurs with the other modelled SuDS techniques [3]. Consequently, as infiltration increased, a slight increase in runoff reduction was achieved for the modelled swales during higher flows associated with more intense storms. Nonetheless, although they had a reduced capacity for detaining runoff, they are a critical component in a SuDS management train [49]. Swales provide open-channel conveyance of runoff in a SuDS management train, which is a more sustainable approach to traditional piped drainage. If designed effectively, for example, as roadside verges, swales enable conveyance of runoff, while also increasing the amenity and biodiversity of a site compared to impermeable surfaces [3,50].

Overall, the SuDS management train is capable of reducing runoff for all modelled rainfall scenarios and at each modelled infiltration type, with a total of 34.86% of the land being attributed to SuDS. It should, however, be noted that the rainfall scenarios modelled were for the 1% AEP storms, as suggested by UK drainage standards and current practice [25]. Consequently, this does not cover the most extreme scenarios, such as 2015 UK storm Desmond, which was calculated to be a 0.1% AEP storm, and also those occurring from a changing climate, which is more likely to cause catastrophic scale flooding [51]. Ellis and Viavattene [52] suggested that during the most extreme rainfall events, SuDS are capable of reducing the scale of the flooding, but are unlikely to fully manage all stormwater, as is the case with the design used for this research and the 1% AEP scenario. However, as previously identified, there is long-held concern regarding the site benefits that can be achieved by integrating SuDS into the drainage scheme, with resistance often as a result of their high land-take [8–10]. The modelled SuDS management train had a net-land take of 7.76%, which could be used for alternative use, if SuDS were not integrated. This is a result of both green roofs (20.34%) and porous pavement (6.76%) being constructed on space that would otherwise be impermeable surfaces, in traditional tiled roofs or impermeable driveways [12]. There are also wider benefits that can be achieved through integrating SuDS, as demonstrated by the SuDS Square, however, focusing on water quantity reductions, simply integrating swales across 3.38% of a site will achieve some reductions in total outflow, with further reductions seen as additional devices are added [3].

5. Conclusions

Whilst research has shown the benefits of integrating SuDS, particularly in a management train, there remain a number of barriers to their wider implementation [6]. This study has presented a novel approach to understand the specific benefits that popular SuDS techniques can have on minimizing peak runoffs, utilizing the UK industry standard drainage modelling tool, MicroDrainage. The results eliminate some of the long-held assumptions regarding the effectiveness of SuDS, and highlight the possible reduction in peak runoff that can be achieved by combining devices, with respect to their land-take. Figures 3–5 show the benefits of installing porous pavement at source control level, and detention basins at the site level, and whilst they take up a combined total of 11.14% of the total modelled site, they retain multiple additional benefits; detention basins can be used for recreational use, and porous pavement can replace existing impermeable surfaces [34,35]. Figure 3d shows that the modelled swales responded differently to changing infiltration rates, compared to other modelled SuDS. As swales are primarily used for conveying stormwater, they are less effective than other techniques as they have a reduced capacity for storage. Consequently, they rely on infiltration rates as their sole method for reducing a limited amount of runoff, therefore, respond differently to the other modelled SuDS.

This study has also highlighted the challenges of using green roofs in MicroDrainage. The underlying Equation (1) for the tool is based upon an assumption that green roofs become saturated after 120 min, and are capable with managing all rainfall up to that duration; therefore, their impact on reducing runoff is far greater for a shorter duration, higher intensity events. This is in contrast to findings by Stovin et al. [40]. It is suggested that whilst it is evident from all scenarios that green roofs are capable of reducing peak runoff to some extent, it is unclear how effective they can be, based on the MicroDrainage model. Validation using physical models is, therefore, required to determine the overall effectiveness of green roofs in the model.

The results establish the capabilities of each of the modelled SuDS devices, reducing the barriers for their uptake by demonstrating the benefits that can be achieved when combined in a SuDS management train, based on the current UK design standards. However, the modelled system accounts for 34.86% of the total site, there is a net-loss of space, in comparison to a traditionally designed site that did not include SuDS, of 7.76%. By carefully selecting the SuDS that are to be used, it is possible to considerably reduce total site outflow at both new build developments and in the existing urban environment, without compromising greatly on space. This investigation, however, has demonstrated the need for future research to better understand how such sites will respond to extreme rainfall events that are expected to be more common in a changing climate.

Author Contributions: Conceptualization, C.L., S.C.; methodology, C.L., F.W. and M.B.; modelling, C.L.; formal analysis, C.L., S.C., F.W. and M.B.; writing—original draft preparation, C.L., and S.C.; writing—review and editing, C.L., S.C., F.W. and M.B.; visualization, C.L., and M.B.; supervision, S.C., F.W. and M.B. All authors have read and agreed to the published version of the manuscript.

Funding: This research received no external funding.

Acknowledgments: The authors would like to thank Coventry University for funding the research, Innovyze for providing a student licence for MicroDrainage and BWB Consulting for technical support with MicroDrainage.

Conflicts of Interest: The authors declare no conflict of interest.

References

1. Dearden, R.A.; Price, S.J. A proposed decision-making framework for a national infiltration SuDS map. *Manag. Environ. Qual. Int. J.* **2012**, *23*, 478–485. [CrossRef]

2. Fletcher, T.D.; Shuster, W.D.; Hunt, W.F.; Ashley, R.; Butler, D.; Arthur, S.; Trowsdale, S.; Barraud, S.; Semádeni-Davies, A.; Bertrand-Krajewski, J.-L.; et al. SUDS, LID, BMPs, WSUD and more—The evolution and application of terminology surrounding urban drainage. *Urban Water J.* **2014**, *12*, 525–542. [CrossRef]

3. Ballard, B.W.; Wilson, S.; Udale-Clarke, H.; Illman, S.; Scott, T.; Ashley, R.; Kellagher, R. *The SuDS Manual (C753) 2015*; CIRIA: London, UK, 2015.

4. Jefferies, C.; Duffy, A.; Berwick, N.; McLean, N.; Hemingway, A. Sustainable Urban Drainage Systems (SUDS) treatment train assessment tool. *Water Sci. Technol.* **2009**, *60*, 1233–1240. [CrossRef]

5. O'Sullivan, J.; Bruen, M.; Purcell, P.J.; Gebre, F. Urban drainage in Ireland—Embracing sustainable systems. *Water Environ. J.* **2011**, *26*, 241–251. [CrossRef]

6. Lashford, C.; Charlesworth, S.; Warwick, F.; Blackett, M. Deconstructing the sustainable drainage management train in terms of water quantity—Preliminary results for Coventry, UK. *CLEAN Soil Air Water* **2014**, *42*, 187–192. [CrossRef]

7. Susdrain. Available online: https://www.susdrain.org/case-studies/ (accessed on 2 February 2020).

8. Melville-Shreeve, P.; Cotterill, S.; Grant, L.; Arahuetes, A.; Stovin, V.; Farmani, R.; Butler, D. State of SuDS delivery in the United Kingdom. *Water Environ. J.* **2017**, *32*, 9–16. [CrossRef]

9. O'Donnell, E.C.; Lamond, J.E.; Thorne, C.R. Recognising barriers to implementation of Blue-Green Infrastructure: A Newcastle case study. *Urban Water J.* **2017**, *14*, 964–971. [CrossRef]

10. Booth, C.; Charlesworth, S. An overture of sustainable surface water management. In *Sustainable Surface Water Management Systems: A Handbook for SuDS*, 1st ed.; Charlesworth, S., Booth, C., Eds.; Wiley Blackwell: London, UK, 2016; Chapter 1; pp. 3–10.

11. Wade, R.; McLean, N. Multiple benefits of green infrastructure. In *Water Resources in the Built Environment: A Handbook for SuDS*, 1st ed.; Charlesworth, S., Booth, C., Eds.; Wiley Blackwell: London, UK, 2016; Chapter 24; pp. 319–335.

12. Jones, P.; Macdonald, N. Making space for unruly water: Sustainable drainage systems and the disciplining of surface runoff. *Geoforum* **2007**, *38*, 534–544. [CrossRef]

13. Stovin, V.; Moore, S.L.; Wall, M.; Ashley, R.M. The potential to retrofit sustainable drainage systems to address combined sewer overflow discharges in the Thames Tideway catchment. *Water Environ. J.* **2012**, *27*, 216–228. [CrossRef]

14. Ministry of Housing, Communities & Local Government. Planning for the Future. 2020. Available online: https://assets.publishing.service.gov.uk/government/uploads/system/uploads/attachment_data/file/872091/Planning_for_the_Future.pdf (accessed on 5 June 2020).

15. Lowe, J.A.; Bernie, D.; Bett, P.; Bricheno, L.; Brown, S.; Calvert, D.; Clark, R.; Eagle, K.; Edwards, T.; Fosser, G.; et al. UKCP18 Science Overview Report. 2018. Available online: https://www.metoffice.gov.uk/pub/data/weather/uk/ukcp18/science-reports/UKCP18-Overview-report.pdf (accessed on 30 August 2020).

16. Sayers, P.; Horritt, M.; Penning-Rowsell, E.; McKenzie, A. Climate Change Risk Assessment 2017 Projections of future flood risk in the UK. 2015. Available online: https://www.theccc.org.uk/wp-content/uploads/2015/10/CCRA-Future-Flooding-Main-Report-Final-06Oct2015.pdf.pdf (accessed on 30 August 2020).

17. Bastien, N.; Arthur, S.; Wallis, S.; Scholz, M. The best management of SuDS treatment trains: A holistic approach. *Water Sci. Technol.* **2010**, *61*, 263–272. [CrossRef]

18. Innovyze. *MicroDrainage v2019.1*; Innovyze: Newbury, UK, 2019.

19. Kellagher, R. *Preliminary Rainfall Runoff Management for Developments Report SC030219*; Environment Agency: Bristol, UK, 2013.

20. Boorman, D.B.; Hollis, J.M.; Lilly, A. *Hydrology of Soil Types: A Hydrologically Based Classification of the Soils of the United Kingdom*; IH Report No. 126; Institute of Hydrology: Wallingford, UK, 1995.

21. Natural Environment Research Council. *Flood Studies Report*; NERC: London, UK, 1975.

22. British Standard Institution (BSI) BS7533–13:2009. *Pavements Constructed with Clay, Natural Stone or Concrete Pavers*; British Standards Institution: London, UK, 2009.

23. Stovin, V. The potential of green roofs to manage Urban Stormwater. *Water Environ. J.* **2009**, *24*, 192–199. [CrossRef]

24. Chow, V.T. *Open-Channel Hydraulics*; McGraw-Hill: New York, NY, USA, 1959.

25. Department for Environment, Food and Rural Affairs. Non-Statutory Technical Standards for Sustainable Drainage Systems. 2015. Available online: https://www.gov.uk/government/uploads/system/uploads/attachment_data/file/415773/sustainable-drainage-technical-standards.pdf (accessed on 3 February 2020).

26. Yazdi, J. Optimal operation of urban storm detention ponds for flood management. *Water Resour. Manag.* **2019**, *33*, 2109–2121. [CrossRef]

27. Scholz, M.; Grabowiecki, P. Review of permeable pavement systems. *Build. Environ.* **2007**, *42*, 3830–3836. [CrossRef]

28. Office for National Statistics. Rural Population and Migration. 2020. Available online: https://assets.publishing.service.gov.uk/government/uploads/system/uploads/attachment_data/file/862320/Rural_population_and_migration_Jan_20.pdf (accessed on 16 April 2020).

29. Miró, A.; Hall, J.; Rae, M.; O'Brien, C.D. Links between ecological and human wealth in drainage ponds in a fast-expanding city, and proposals for design and management. *Landsc. Urban Plan.* **2018**, *180*, 93–102. [CrossRef]

30. Gaborit, E.; Muschalla, D.; Vallet, B.; Vanrolleghem, P.; Anctil, F. Improving the performance of stormwater detention basins by real-time control using rainfall forecasts. *Urban Water J.* **2013**, *10*, 230–246. [CrossRef]

31. Potter, K.; Vilcan, T. Managing urban flood resilience through the English planning system: Insights from the 'SuDS-face'. *Philos. Trans. R. Soc. A Math. Phys. Eng. Sci.* **2020**, *378*, 20190206. [CrossRef]

32. Fenner, R.A.; Andrew, R.F. Spatial evaluation of multiple benefits to encourage multi-functional design of sustainable drainage in blue-green cities. *Water* **2017**, *9*, 953. [CrossRef]

33. Ellis, J.B.; Lundy, L. Implementing sustainable drainage systems for urban surface water management within the regulatory framework in England and Wales. *J. Environ. Manag.* **2016**, *183*, 630–636. [CrossRef] [PubMed]

34. Wihlborg, M.; Sörensen, J.; Olsson, J.A. Assessment of barriers and drivers for implementation of blue-green solutions in Swedish municipalities. *J. Environ. Manag.* **2019**, *233*, 706–718. [CrossRef]

35. Backhaus, A.; Dam, T.; Jensen, M.B. Stormwater management challenges as revealed through a design experiment with professional landscape architects. *Urban Water J.* **2012**, *9*, 29–43. [CrossRef]

36. Tsavdaris, A.; Mitchell, S.; Williams, J.B. Computational fluid dynamics modelling of different detention pond configurations in the interest of sustainable flow regimes and gravity sedimentation potential. *Water Environ. J.* **2014**, *29*, 129–139. [CrossRef]

37. Miller, J.D.; Hutchins, M. The impacts of urbanisation and climate change on urban flooding and urban water quality: A review of the evidence concerning the United Kingdom. *J. Hydrol. Reg. Stud.* **2017**, *12*, 345–362. [CrossRef]

38. Chen, J.; Liu, Y.; Gitau, M.W.; Engel, B.A.; Flanagan, D.C.; Harbor, J.M. Evaluation of the effectiveness of green infrastructure on hydrology and water quality in a combined sewer overflow community. *Sci. Total. Environ.* **2019**, *665*, 69–79. [CrossRef] [PubMed]

39. Gordon-Walker, S.; Harle, T.; Naismith, I. Cost-Benefit of SUDS Retrofit in Urban Areas. 2007. Available online: https://assets.publishing.service.gov.uk/government/uploads/system/uploads/attachment_data/file/290993/scho0408bnxz-e-e.pdf (accessed on 3 February 2020).

40. Stovin, V.; Vesuviano, G.; Kasmin, H. The hydrological performance of a green roof test bed under UK climatic conditions. *J. Hydrol.* **2012**, *414*, 148–161. [CrossRef]

41. Berndtsson, J.C. Green roof performance towards management of runoff water quantity and quality: A review. *Ecol. Eng.* **2010**, *36*, 351–360. [CrossRef]

42. De-Ville, S.; Menon, M.; Jia, X.; Reed, G.; Stovin, V. The impact of green roof ageing on substrate characteristics and hydrological performance. *J. Hydrol.* **2017**, *547*, 332–344. [CrossRef]

43. Charlesworth, S. A review of the adaptation and mitigation of global climate change using sustainable drainage in cities. *J. Water Clim. Chang.* **2010**, *1*, 165–180. [CrossRef]

44. Hoang, L.; Fenner, R.A. System interactions of stormwater management using sustainable urban drainage systems and green infrastructure. *Urban Water J.* **2015**, *13*, 739–758. [CrossRef]

45. Davis, A.P.; Stagge, J.H.; Jamil, E.; Kim, H. Hydraulic performance of grass swales for managing highway runoff. *Water Res.* **2012**, *46*, 6775–6786. [CrossRef]

46. Winston, R.J.; Powell, J.T.; Hunt, W.F. Retrofitting a grass swale with rock check dams: Hydrologic impacts. *Urban Water J.* **2018**, *16*, 404–411. [CrossRef]

47. Allen, D.; Olive, V.; Arthur, S.; Haynes, H. Urban sediment transport through an established vegetated swale: Long term treatment efficiencies and deposition. *Water* **2015**, *7*, 1046–1067. [CrossRef]

48. Fach, S.; Engelhard, C.; Wittke, N.; Rauch, W. Performance of infiltration swales with regard to operation in winter times in an Alpine region. *Water Sci. Technol.* **2011**, *63*, 2658–2665. [CrossRef] [PubMed]

49. Woznicki, S.A.; Hondula, K.L.; Jarnagin, S.T. Effectiveness of landscape-based green infrastructure for stormwater management in suburban catchments. *Hydrol. Process.* **2018**, *32*, 2346–2361. [CrossRef]

50. García-Serrana, M.; Gulliver, J.S.; Nieber, J.L. Non-uniform overland flow-infiltration model for roadside swales. *J. Hydrol.* **2017**, *552*, 586–599. [CrossRef]

51. Marsh, T.; Kirby, C.; Muchan, K.; Barker, L.; Henderson, E.; Hannaford, J. *The Winter Floods of 2015/2016 in the UK—A Review*; Centre for Ecology and Hydrology: Wallingford, UK, 2016.

52. Ellis, J.B.; Viavattene, C. Sustainable Urban drainage system modeling for managing urban surface water flood risk. *CLEAN Soil Air Water* **2013**, *42*, 153–159. [CrossRef]

Article

Water Quality Control Options in Response to Catchment Urbanization: A Scenario Analysis by SWAT

Hong Hanh Nguyen *, Friedrich Recknagel and Wayne Meyer

Department of Ecology and Environmental Sciences, University of Adelaide, Adelaide SA 5005, Australia;
friedrich.recknagel@adelaide.edu.au (F.R.); wayne.meyer@adelaide.edu.au (W.M.)
* Correspondence: hanh.nguyen@adelaide.edu.au; Tel.: +61-044-919-2658

Received: 7 November 2018; Accepted: 11 December 2018; Published: 13 December 2018

Abstract: Urbanization poses a challenge to sustainable catchment management worldwide. This study compares streamflows and nutrient loads in the urbanized Torrens catchment in South Australia at present and future urbanization levels, and addresses possible mitigation of urbanization effects by means of the control measures: river bank stabilization, buffer strip expansion, and wetland construction. A scenario analysis by means of the Soil and Water Assessment Tool (SWAT) based on the anticipated urban population density growth in the Torrens catchment over the next 30 years predicted a remarkable increase of streamflow and Total Phosphorous loads but decreased Total Nitrogen loads. In contrast, minor changes of model outputs were predicted under the present urbanization scenario, i.e. urban area expansion on the grassland. Scenarios of three feasible control measures demonstrated best results for expanding buffer zone to sustain stream water quality. The construction of wetlands along the Torrens River resulted in the reduction of catchment runoff, but only slight decreases in TN and TP loads. Overall, the results of this study suggested that combining the three best management practices by the adaptive development of buffer zones, wetlands and stabilized river banks might help to control efficiently the increased run-off and TP loads by the projected urbanization of the River Torrens catchment.

Keywords: SWAT; urbanization; nutrient loads; constructed wetlands; buffer zones; river bank stabilization

---

1. Introduction

Urbanization is the most common trend in land use changes worldwide, with approximately half of the global population residing in disproportionately small areas of land [1–3]. The urbanization of catchments is associated with sealing, compaction, degradation, and mixing of natural soils with imported soils [4,5], and requires informed sustainable management. Increased runoff and erosion rates, degraded water quality, reduction in biodiversity, wetland loss, and eutrophication are some of the consequences of rapid urbanization [6,7]. Analysis of 106 river catchments worldwide found that the proportion of catchments with streamflow being fragmented and disturbed by dams in urban areas is projected to increase to 70% by 2050 [7]. In Australia, natural catchments have been drastically altered since European settlement by land clearing and development of cities. Approximately 90% of the Australian population is living in urban areas [1] and many catchments face the risk of elevated nutrient loads and substantial algal blooms [8–10]. Thus, studies that allow quantitative evaluation of effects of urbanization are of great importance for the future water-sensitive design of Australian cities [11]. Catchment modeling has been defined as an important tool to assist this target [11,12].

The Soil and Water Assessment Tool (SWAT) is a widely used catchment modeling tool that allows to predict streamflow and non-point source pollutants under varying soil, land use and management conditions worldwide [13–15]. The SWAT model was originally developed to simulate rural catchments, but algorithms describing urban processes were later incorporated in the model [15]. Results from many studies by the SWAT have suggested that urbanization causes significant alterations in the water budget of catchments by increasing surface runoff and decreasing baseflow in streams [16–20]. Some studies have also reported a linear relationship between the speed of urbanization and the increase in sediment and nutrient loads [21,22]. In the study by Lee et al. [23], the projection of urbanization for 2030 suggests increases of total nitrogen and total phosphorous in many catchments by up to 24% and 111%, respectively. As stated by Wang and Kalin [22], substantial urbanization on forest lands is expected to cause higher peaks for sediment and total phosphorous loads during wet seasons, whereas rapid urbanization may even have a stronger effect on nitrogen and phosphorous than projected climate change [24]. In the case of Australia, most studies on catchment urbanization have focused on hydrological impacts [25] whilst studies on nutrient loads have been applied to agricultural catchments [26–28].

The River Torrens catchment covers an area of 200 km^2 and is located in the central part of Adelaide, the capital of South Australia. It supplies drinking water, environmental flow, and fulfills recreational and conservational purposes for the capital city [29]. Urban development is affecting water quality of its tributaries and creeks [8].

This study focuses on modelling effects of urbanization on streamflow and nutrient loads in the Torrens catchment using the SWAT model. The study also examines the effectiveness of potential mitigating control options in response to future catchment urbanization, and improves understanding of this issue for urban catchment managers and policy makers. This case study may also of interest to modelers working on similar environmental problems around the world.

2. Materials and Methods

2.1. Study Area

The urban section of the River Torrens catchment below Gorge weir (hereafter called the Urban Torrens catchment) was used throughout this study for the SWAT model application. The study area includes the First to Fifth Creek and the River Torrens, which pass through the Adelaide Central Business District (CBD). The catchment lies between latitude −34°51′23″ and −34°56′53″ S and longitude 138°32′55″ to 138°43′52″ E. The altitude of this area ranges from 9 to 681 m with an average value of 214 m. The Mediterranean climate of the study area is characterized by a low average annual rainfall of 600 mm that is mostly concentrated in sporadic storm events in summer or during the wet winter.

2.2. Input Data

2.2.1. Soil Data

The soil inputs required for the SWAT model comprise of soil maps and soil attribute data (Figure 1). The soil maps of the study area include the map of South Australia, which was provided by the South Australian Department of Agriculture, and a map of the urban area which was extracted from a project on mapping soils around metropolitan Adelaide by the Department of State Development [30]. This project was carried out to explore the properties of Metropolitan Adelaide soils, which include some reactive soils and clays that are sensitive to seasonal and human-induced changes and have caused severe failure of masonry buildings in many urban regions around Adelaide. Both maps were provided at the resolution of 1:100,000 and were clipped to prepare unique raster map using a geographic information system (GIS) tool.

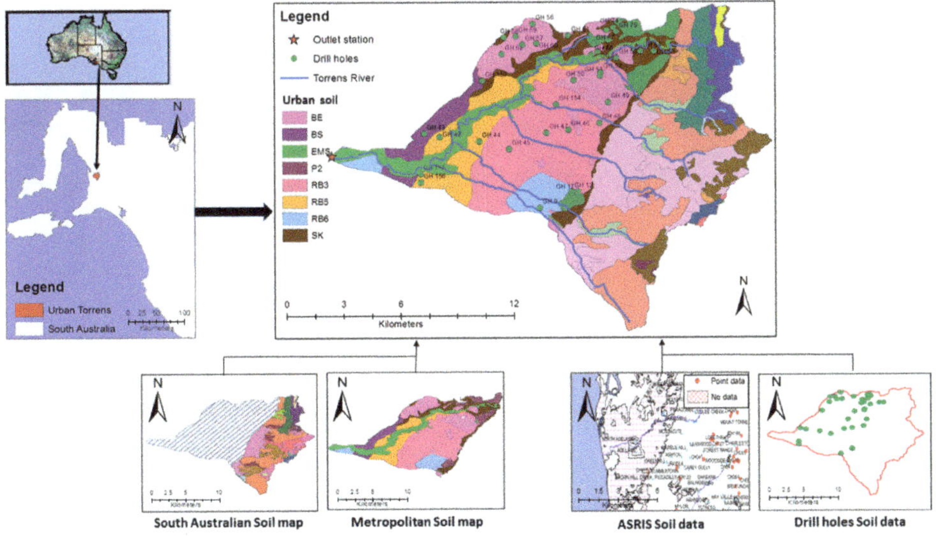

Figure 1. Soil maps of the study area.

For the SWAT database, the major soil information was provided by the Australian Soil Resource Information System (ASRIS) [31], while the attributes of soils in the missing information area (Figure 1) were constructed on the basis of data available from the Drill Core Reference Library, published literature [30] and expert knowledge. Information from 27 data points drilled to 10 m depth [28] was combined to develop eight major soil classes: black earth (BE), brown solonized (BS), estuarine sediments (EM), podzolic (P2), red-brown earths (RB3, RB5, RB6), and solodic (SK) soils. These soil classes were further characterized by soil attributes comprised of soil layers, soil hydrological groups, plant root depth, soil dry bulk density, soil organic content, and percent of clay, silt, sand, and rock fragments. Some soil parameters were estimated using the following functions [32,33]:

$$\theta_p = 0.132 - 2.5 \times 10^{-6} \times e^{0.105 \times \%sand} \tag{1}$$

$$K_{sat} = 750 \times \left(\frac{\theta_{sat} - \theta_d}{\theta_d} \right)^2 \tag{2}$$

where θ_p (m³ H$_2$O/m³ soil) is the soil available water content (SOL_AWC) (mm H$_2$O/mm soil), K_{sat} is the saturated hydraulic conductivity (SOL_K) (mm/day), θ_{sat} is the upper limit of water content that is possible in a soil of known bulk density and θ_d (m³ H$_2$O/m³ soil) is the volumetric drained upper limit water content.

θ_{sat} was calculated from Soil bulk density (ρ_d) (g/cm³) using the following formula:

$$\theta_{sat} = 1.0 - \frac{\rho_d}{2.65} \tag{3}$$

θ_d was calculated from the gravimetric drained upper limit w_d (kg H$_2$O/kg Soil) and ρ_d:

$$\theta_d = w_d \times \rho_d \tag{4}$$

$$w_d = 0.186 \times \left(\frac{sand}{clay} \right)^{-0.141} \tag{5}$$

These equations have been successfully applied to derive soil characteristics for a range of soils in south eastern South Australia [34]. The soil erodibility (USLE_K) parameter (0.013 (metric ton m^2 h)/(m^3-metric ton cm)) was estimated from relative proportions of sand, silt, and clay in each soil layer using the method provided in the SWAT manual [15].

The resulting attributes of the soil profile of the Urban Torrens catchment are provided in Table 1. These include average data for two soil layers of soil classed which are estimated from drill hole information and more detailed data on five soil layers of soil classes provided by the ASRIS source.

Table 1. Characteristics of soil database in the Urban Torrens catchment.

Soil Profile	Layer (s)	Soil Parameters *										Data Source
		SOL_Z	SOL_BD	SOL_AWC	SOL_K	SOL_CBN	CLAY	SILT	SAND	SOL_ALB	USLE_K	
	1	975	1.379	0.131	21.6	2.25	43	27	30	0.17	0.051	Drill holes [30]
	2	4071	1.389	0.132	23.3	0.50	43	30	27	0.17	0.051	
	1	126	1.403	0.129	31.0	2.76	16	18	66	0.18	0.051	
	2	217	1.503	0.126	20.7	0.73	16	13	71	0.18	0.051	
	3	349	1.428	0.132	7.0	0.49	43	37	20	0.18	0.058	
	4	498	1.437	0.132	7.1	0.25	38	36	25	0.18	0.058	ASRIS [31]
	5	876	1.344	0.130	4.8	0.14	26	25	38	0.18	0.050	

Note: * Soil parameters: SOL_Z: soil depth (mm); SOL_BD: moist bulk density (mg/m^3); SOL_AWC: available water capacity (mm/mm soil); SOL_K: saturated hydraulic conductivity (mm/h); SOL_CBN: organic carbon content (%); CLAY: clay content (%); SILT: silt content (%); SAND: sand content (%); SOL_ALB: moist albedo; USLE_K: Universal Soil Loss Equation (USLE) equation soil erodibility (K) factor (0.013 (metric ton m^2 h)/(m^3-metric ton cm)).

2.2.2. Other Input Data

In addition to soil data, application of the SWAT to the Urban Torrens catchment requires a number of input data types and maps:

- Digital elevation model (DEM): the 10 m resolution DEM was interpolated from a 10 m contour map provided by the SA Water Corporation.
- Flow burn-in layer: The river network was superimposed onto the DEM to adjust the location of some downstream urban creeks that were not well predicted by DEM due to modification effects from urban land development. The burn-in river layer was provided by the SA Water Corporation.
- Land use maps: a historical land use map at a scale of 1:100,000, which was completed in 2007 and updated with recent data on locations and land uses of the Torrens catchment, was provided by the Department of Environment, Water and Natural Resources. The map classifies the catchment into urban residential, commercial, institutional, industrial, transportation, water, and grassland land uses. For the past land use scenario, a historical map of 2001 of the whole South Australia was provided by the Department of Planning, Transport and Infrastructure.
- Climate data: this includes maximum and minimum air temperature, rainfall, relative humidity, and solar radiation. The daily data for these variables from 2008 to 2015 from five weather stations was extracted from the Scientific Information for Land Owners (SILO) website [35].
- Streamflow and nutrient data: data of daily streamflow and monthly composite Total Nitrogen (TN) and Total Phosphorous (TP) loads at the outlet of the study area (Holbrooks Road Station, A5040529) were provided by the Adelaide and Mount Lofty Ranges Natural Resources Management Board [36]. Data were extracted for the period from 2008 to 2015.

2.3. Soil and Water Assessment Tool (SWAT) Model Set-Up

SWAT (ArcSWAT version 2012 revision 637, USDA, Washington, DC, USA) is a continuous-time, semi-distributed simulator developed to assist water resource managers in predicting impacts of land management practices on water quality, including various species of nitrogen and phosphorous [13,15]. Spatially, the model subdivides a catchment into sub-basins, which are further delineated into hydrological response units (HRUs) based on physical characteristics of topography, soil, and land uses. In this study, application of the SWAT model resulted in a subdivision of the Urban Torrens catchment into 23 sub-basins and further into 125 HRUs using the multiple HRU thresholds method of soil, land use, and slope at 10, 20, and 10%, respectively. A modified Soil Conservation Service (SCS) curve number technique was used to estimate the streamflow, while the instream processes of TN and TP loads were estimated using the Enhanced Stream Water Quality Model (QUAL2E) [37]. Local information on management practices was imported into the model on the basis of expert knowledge. All land operations were scheduled by specific application date [15]. The growing season was defined from 1 June to 30 May for all urban land categories. In order to simulate management activities along land uses by agriculture, pasture, and orchards, the approach designed by Nguyen et al. [28] has been applied.

The parameter optimization of the SWAT model was based on sensitivity analysis, model calibration, model validation, and uncertainty analysis. These steps are in accordance with Neitsch et al. [15] and Arnold et al. [38], and will be discussed in the following section.

2.3.1. Parameter Sensitivity Analysis

The sequential uncertainty fitting (SUFI2) algorithm [38] of the SWAT Calibration and Uncertainty Program (SWAT-CUP, EAWAG, Dübendorf, Switzerland) allows analysis of global and one-at-a-time sensitivity. Here we applied the global sensitivity analysis to identify parameters for the calibration and validation steps.

2.3.2. Model Calibration, Validation and Uncertainty

The parameter optimization was performed on a monthly time step using the generalized likelihood uncertainty (GLUE) algorithm that showed better calibration results for this case study when compared to the results of the SUFI2 program. GLUE performs a combined calibration and uncertainty analysis and accounts for all sources of uncertainties [39–41]. The calibration was conducted consecutively beginning with the streamflow followed by loads of sediment (TSS), TN and TP by means of the observed data from 2008 to 2015 and using the first three years as a warm-up period for model stabilization. Data from 2011 to 2013 were used for calibration, and validation was performed for the years 2014 and 2015. 5000 iterations were applied and the Nash-Sutcliffe (NS) [42] behavioral threshold of 0.5 was used for both streamflow and nutrient simulations. The coefficient of determination (R^2), percent bias (PBIAS) [43], and NS efficiency coefficient were used as statistical criteria for evaluation of simulated results.

$$\mathrm{NS} = 1 - \frac{\sum_i \left(Q_{m,i} - Q_{s,i}\right)^2}{\sum_i \left(Q_{m,i} - \overline{Q_m}\right)^2} \tag{6}$$

where: Q is the streamflow variable, m and s are measured and simulated values respectively, i is the ith datum, and the bar stands for average values.

The threshold for R^2 and NS greater than 0.5 for streamflow, TN and TP loads, and PBIAS ranging between ±25% for streamflow and ±70% for TN and TP loads, respectively, were considered as satisfactory modelling results [44]. The model uncertainty was expressed using the 95% prediction uncertainty index (95PPU) and statistically was evaluated based on the percentage of observation points bracketed by the prediction uncertainty band (p-factor) and the degree of uncertainty (r-factor). The values close to 1 were selected as satisfactory criteria for p- and r-factors [45].

2.4. Scenario Analysis

The calibrated model was used to simulate present and future scenarios of urbanization, and determine best-management practices (BMPs). The past (P) and present (BS) urbanization scenarios were represented through land use maps generated in ArcGIS, which indicated a substantial shift in the period from 2001 to 2015 from grassland to urban lands of low residential, institutional, and commercial lands (Figure 2). For the future urbanization scenario (FS0), the urban land budget will not change significantly according to the 'The 30 year Plan for Greater Adelaide' report, even though the urban population density is expected to triple [46]. Therefore, we maintained the relative percentage of land uses from 2015 (Figure 2), and reclassified the land use from low residential to high residential. The change in residential land use was reflected by an increase in the fraction of total impervious areas (FIMP) from 0.12 to 0.6, the amount of solids allowed to build up on impervious area (DIRTMX) from 125 to 225 kg/curb km, TN concentration in suspended solid loads from impervious area (TNCONC) from 360 to 550 mg N/kg sediment, and TP concentration in suspended solid loads from impervious area (TPCONC) from 96 to 223 mg P/kg sediment [15]. Values of parameters for the high-residential land use were extracted from the default database, while data for the low-residential land use were manually calibrated prior the auto-calibration step [47]. Meteorological input data were kept unchanged for all urbanization scenarios.

In order to determine potential BMPs for mediating water deterioration issues by urbanization, the following scenarios were designed:

- Scenario 'Stream bank stabilization' (S1) was set up by increasing vegetative cover (CH_COV2) and Manning's stream roughness coefficient (CH_N2), and reducing the stream erosion (CH_EROD) values by 50% [48–50].
- Scenario 'Buffer strip application' (S2) was set up by extending the 30-m width of the filter strip of alfa grass along the main river using the FILTERW parameter in SWAT '.mgt' input file [51].

- Scenario 'Wetland development' (S3) was represented by a wetland with a maximum surface of 3445 m^2 and volume of 3700 m^3 in the '.pnd' input file, as suggested by Kasan [52]. The nitrogen and phosphorous settling rates were set to 20 m/year using the maximum default value in the '.pnd' input file for systems with high removal efficiency [38]. The bottom hydraulic conductivity was set at 2.3 mm/h [53], and sediment concentration in the wetland was defined at 10 mg/L. The same parameter values were applied to all wetland scenarios of this study.
- Combined scenario (Sm) which simulated together the three aforementioned scenarios.

Results for the past and future urbanization scenarios (P and FS) were compared with results of the present urbanization scenario BS. Results of the scenarios S1, S2 and S3 were compared with the scenario FS0. The statistical significance of scenarios of urbanization and BMPs were evaluated by means of a paired Wilcoxon test using an R tool according to the criteria $p < 0.05$.

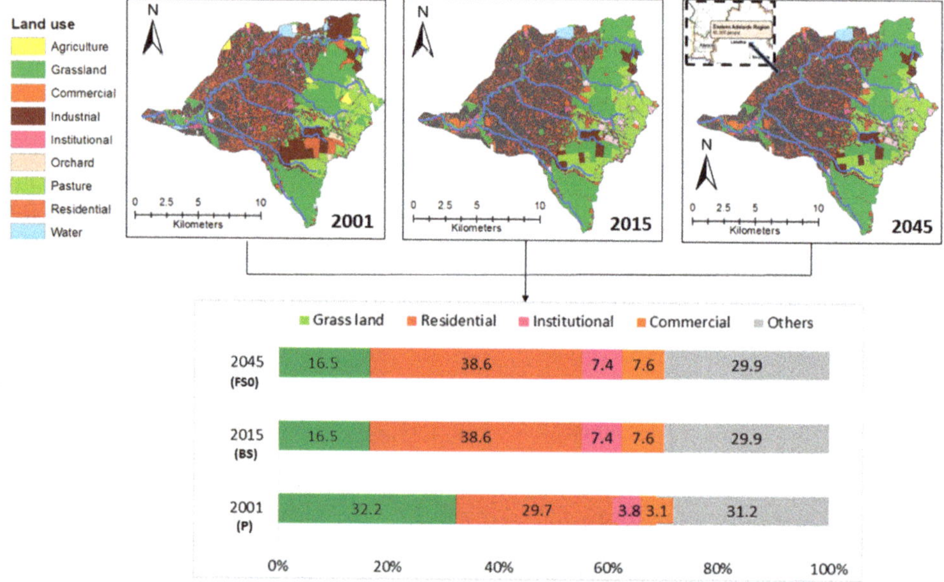

Figure 2. Characteristics of past (P), present (BS), and future (FS0) land use changes in the Urban Torrens catchment from 2001 to 2045.

3. Results

3.1. Model Sensitivity

The global sensitivity analysis identified the runoff curve number (CN2), the baseflow alpha factor for bank storage (ALPHA_BNK) and the moist bulk density (SOL_BD) as most sensitive parameters for streamflow simulation whereas soil parameters SOL_BD, SOL_K, and SOL_AWC were amongst the 10 most sensitive parameters (Table 2). In contrast, the organic N in the baseflow (LAT_ORGN), the denitrification exponential rate coefficient (CDN) and denitrification threshold water content (SDNCO) proved most sensitive parameters for TN-load.

Table 2. Soil and Water Assessment Tool (SWAT) parameters used for model calibration.

Parameters	Description	Unit	Fitted Value	Parameter Sensitivity		
				t-Stat	*p*-Value	Rank
Streamflow						
CN2.mgt	Moisture condition II runoff curve number	-	-0.25 [b]	-63.56	0.00	1
ALPHA_BNK.rte	Baseflow alpha factor for bank storage	-	0.72	29.20	0.00	2
SOL_BD (1,2) [a].sol	Moist bulk density	g/cm^3	-0.19 [b]	-24.50	0.00	3
GWQMN.gw	Threshold depth of water in the shallow aquifer required for return flow to occur	mm H_2O	1854	17.00	0.00	4
ESCO.hru	Soil evaporation compensation factor	-	0.75	-11.91	0.00	5
SOL_K (1,2) [a].sol	Saturated hydraulic conductivity	mm/h	-0.17 [b]	-8.66	0.00	6
SOL_AWC (1,2) [a]	Available water capacity of the soil layer	mm H_2O/mm soil	-0.02 [b]	7.40	0.00	7
CH_K2.rte	Effective hydraulic conductivity in main channel alluvium	mm/h	59.6	-7.08	0.00	8
CH_N2.rte	Manning's "n" value for the main channel	-	0.04	-5.98	0.00	9
GW_REVAP.gw	Groundwater "revap" coefficient	mm H_2O	0.19	3.66	0.00	10
GW_DELAY.gw	Groundwater delay	days	221.3	3.51	0.00	11
RCHRG_DP.gw	Deep aquifer percolation fraction	-	0.17	-2.94	0.00	12
Total Suspended Solid Load						
USLE_P.mgt	USLE equation support practice factor	-	0.39	-63.26	0.00	1
CH_COV1.rte	Channel erodibility factor	-	0.32	1.98	0.05	2
SPEXP.bsn	Exponent parameter for calculating sediment re-entrained in channel sediment routing	-	1.12	1.76	0.08	3
CH_EROD.rte	Channel erodibility factor	-	0.56	1.30	0.19	4
SPCON.bsn	Linear parameter for calculating the maximum amount of sediment that can be reentrained during channel sediment routing	-	0.006	0.38	0.70	5
CH_COV2.rte	Channel cover factor	-	0.62	0.04	0.97	6
Total Nitrogen Load						
LAT_ORGN.gw	Organic N in the baseflow	mg/L	6.33	-167.44	0.00	1
CDN.bsn	Denitrification exponential rate coefficient	-	0.56	-7.49	0.00	2
SDNCO.bsn	Denitrification threshold water content	-	0.73	3.8	0.00	3
ERORGN.hru	Organic nitrogen enrichment ratio	-	1.27	-1.08	0.28	4
NPERCO.bsn	Nitrogen percolation coefficient	-	0.15	-0.23	0.82	5
Total Phosphorous Load						
PHOSKD.bsn	Phosphorus soil partitioning coefficient	-	187.03	-0.88	0.38	1
PSP.bsn	Phosphorus sorption coefficient	-	0.06	-0.78	0.43	2
ERORGP.hru	Organic phosphorus enrichment ratio	-	2.51	0.49	0.62	3

Note: [a] Values in parentheses indicate the soil layer; [b] Indicated value refers to a relative change in the parameter.

3.2. Model Calibration, Validation and Uncertainty

Calibrations for streamflow, TN, and TP resulted in coefficients of determination R^2 of 0.77, 0.62, and 0.56, NS of 0.77, 0.62, and 0.51, and PBIAS of -4.18, -2.91, and 24.87 respectively (Figure 3) that according to Moriasi et al. [44] indicate to be satisfactory. Validation for streamflow achieved $R^2 = 0.97$, NS = 0.96, and PBIAS = -9.21, for TP $R^2 = 0.88$, NS = 0.84, and PBIAS = -28.4 and for TN with $R^2 = 0.67$, NS = 0.66, and PBIAS = -2.60. The p-factor for the uncertainty for flow ranged between 0.39 and 0.42, for TN between 0.83 and 0.71, for TP between 0.56 and 0.54, and the r-factor ranged between 0.75 and 0.79 for flow, 1.32 and 0.96 for TN, and 1.00 and 0.83 for TP during calibration and validation, respectively. The simulated peaks of streamflow, TN and TP loads corresponded well with monthly average precipitation in this urbanized catchment.

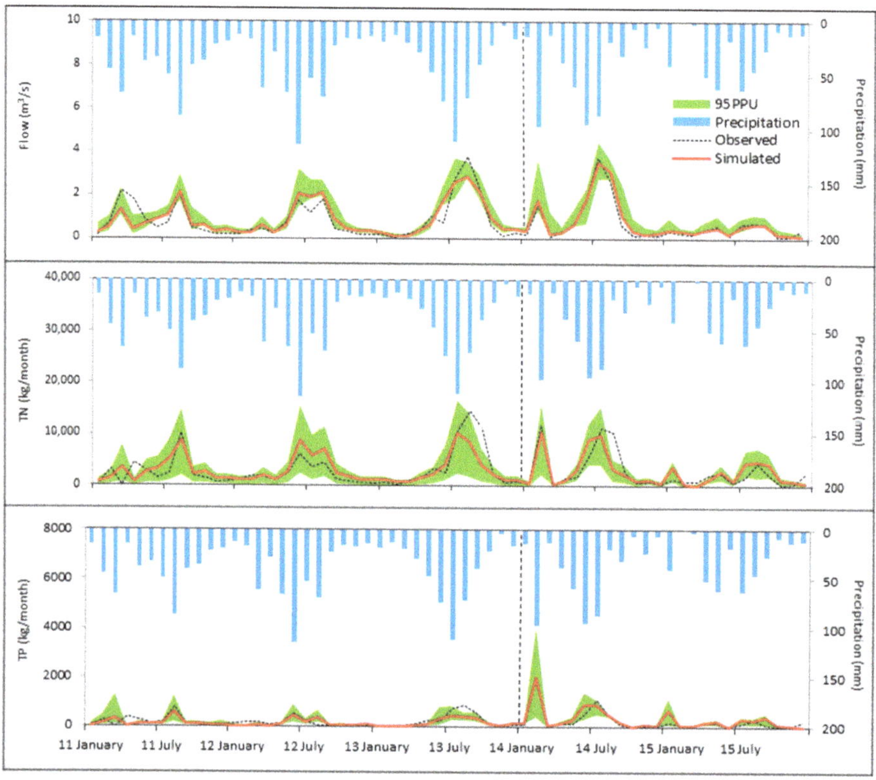

Figure 3. Hydrographs of observed and simulated streamflow and TN and TP loads of the Urban Torrens catchment during the calibration (2011–2013) and validation (2014–2015) periods.

3.3. Urbanization Scenarios

Results of the scenario BS indicated an overall increase of 0.6% in monthly streamflow due to an increase of surface streamflow by 1% and a decrease of baseflow by approximately 2% (Figure 4). Whilst scenario BS also predicted an increase of the TP load by the 2.9% forecasted, TN loads changed insignificantly compared to the past urbanization scenario. The trends for streamflow, TN and TP are relatively similar for all months of the year. The scenario FS0 (future urbanization) suggested a significant increase in total runoff by 13.3% when compared to present urbanization. The partitioning of streamflow under the scenario FS0 (Figure 4) indicates a similar trend of an increasing surface streamflow from 77 to 82%, while baseflow is further decreasing from 23 to 18%. There is also a significant increase by 36.4% of the TP-loads at the catchment outlet suggested. Meanwhile, model results suggest a noticeable decrease in TN loads of 6.9%. From the results of the future urbanization scenario it is also evident that higher rates of nutrient load variations are observed for the rainy period in winter (June to August). Overall, the trend is clear and similar when the effects of past, present and future urbanization scenarios are compared with more pronounced effects of future urbanization versu s present urbanization.

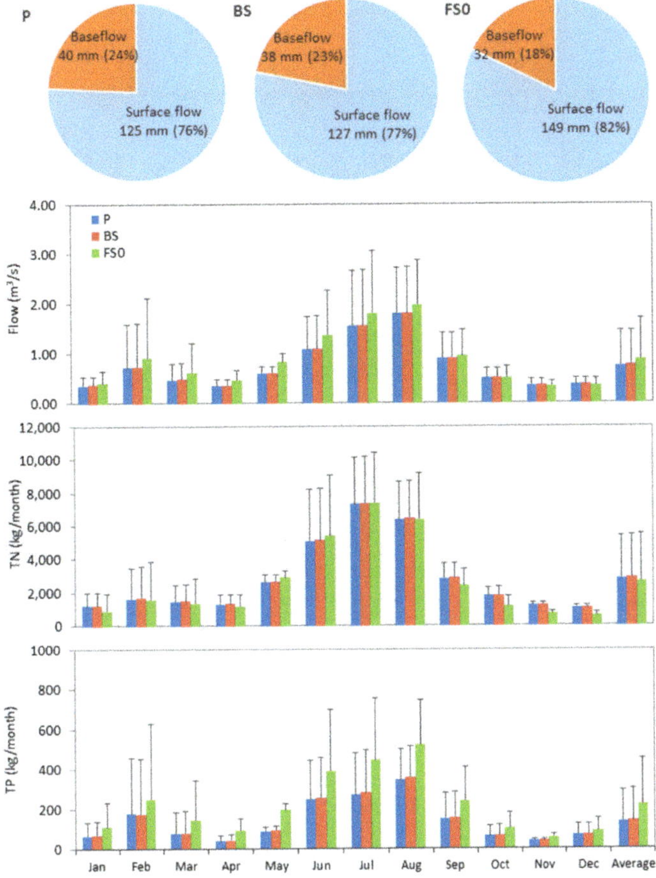

Figure 4. Streamflow, TN and TP responses to scenarios past urbanisation (P), present urbanisation (BS) and future urbanization (FSO). Pie charts show the relative proportion of different hydrological components. Bar graphs show the average streamflow, TN, and TP loads. Error bars show one standard deviation.

3.4. Scenarios of Management Practices

The Table 3 suggests that the scenario '30-m buffer strips' may achieve the highest reduction of the TN loads by 19.88% and of the TP loads by 4.13% compared to 1.22% and 2.73%, respectively, by the scenario 'river bank stabilization'. However, both scenarios predicted statistically insignificant changes in the catchment outflow. The scenario 'wetland development' showed a slight decrease in TN and TP loads, and buffering effects for the increased run-off into the main stream. The scenario that combined the three feasible management practices predicted a decreased runoff and the highest reduction in nutrient loads compared to results of the scenarios of the three single measures.

Table 3. Results of best management scenarios for flow, TN and TP loads at the Urban Torrens catchment. The relative change of best-management practices (BMP) scenarios are compared with the results of the FS0 scenario.

Scenarios	Flow		TN Load		TP Load	
	Mean Values (m³/s)	Relative Change (%)	Mean Values (tons/year)	Relative Change (%)	Mean Values (tons/year)	Relative Change (%)
River bank stabilization—S1	0.88	<1	31.65 [a]	−1.22	2.57 [a]	−2.73
30-m buffer strips—S2	0.88	<1	25.67 [a]	−19.88	2.53 [a]	−4.13
Wetland development—S3	0.86 [a]	−2.27	31.76 [a]	−0.87	2.58 [a]	−2.44
Combined BMPs—Sm	0.86 [a]	−2.28	25.21 [a]	−21.30	2.47 [a]	−6.40

[a] indicates a significant different value (p-value < 0.05) for a BMP scenario as compared with the FS0 scenario based on the paired Wilcoxon test.

4. Discussion

This study applied SWAT for modelling impacts of urbanization on the Torrens catchment that is of high relevance Australia-wide.

With regards to model optimization, it proved to be advantageous to include field-based soil database of the Torrens catchment as model input that resulted in satisfactory streamflow simulation of both peak and base flows (Figures 3 and 5) and improved simulation results for nutrient loads when compared with results for the urban catchment Aldgate of a previous study by Shrestha et al. [27] that was based on a coarser representation of soils.

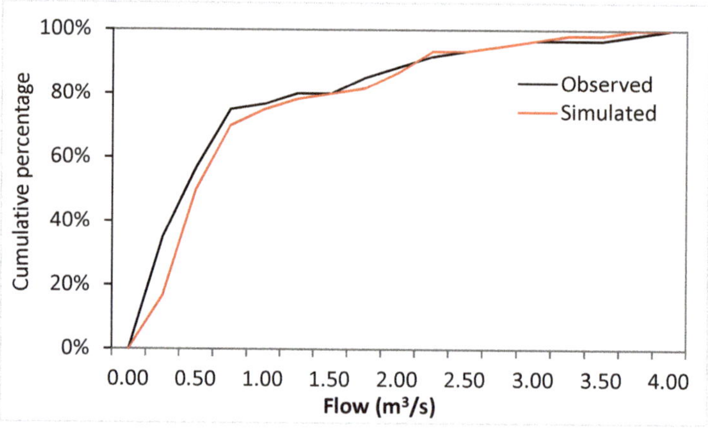

Figure 5. Flow duration curve of observed and simulated streamflow of the Urban Torrens catchment for the period from 2011 to 2015.

All urbanization-related scenarios predicted increased streamflow as a result of increased surface flow and decreased baseflow that corresponds well with findings by Richards et al. [16] and Sunde et al. [19]. The trends of predicted TP loads as appeared to be strongly positively correlated with streamflow since phosphorus is primarily transported by sediments in surface streamflow. The model predicted annual increases of TP loads by 4 g/ha/year in scenario BS and 65 g/ha/year in scenario FS0. In contrast, the scenario results showed that urbanization may decrease TN load most likely because of reduced soil leaching by up to 26 g/ha/year and up to 2 g/ha/year less nitrogen in the baseflow as revealed by the comparison between the scenarios FS0 and BS. The highest changes in nutrient loads were recorded during autumn and winter months when pollutants are often released and transported in river catchments during short periods of intensive rainfall [8,54].

The comparison between the scenarios P and FS0 revealed significant increases in streamflow by 13.3% and in TP loads by 36.4% whilst TN-loads decreased by 6.9%. A possible explanation lies in the fact that pervious urban lands are modelled in SWAT as Bermuda grass, which in this study is configured similarly to pasture and grassland. Thus, the conversion of the low-residential land use accounting for 38.6% of the total land budget of scenario P to high-residential land by scenario FS0 corresponded to an increase of overall impervious surface in the study by approximately 20%. According to the study of Brun and Band [55], 20% is the threshold at which a dramatic change in runoff can be observed. It is also important to mention that in the case of the Urban Torrens catchment, the sewage system is completely separated from the stormwater drainage network. Therefore, an increasing urban population is projected to cause more fragmented housing sites and smaller-sized yards but not necessarily an increase in surface flow by waste water, and simulated streamflows and nutrient loads are only driven by stormwater.

In an attempt to determine measures for counteracting the impacts of urbanization we have examined three management options. The scenario that simulated the extension of the grassed buffer zone proved to be efficient in reducing TP loads whilst developing wetlands may buffer the flow into the main rivers. However, the implementation of these two measures in combination with river bank stabilization promises to be the best management practice in response to future urbanization of the Urban Torrens catchment.

5. Conclusions

As outcomes of this study, the following conclusions were drawn:

- Growing urbanization increases surface flow and TP loads whereas baseflow and TN loads decrease due to extending impervious area.
- Expanded buffer zones and stabilized river banks can retain nutrients while constructing adjacent wetlands may reduce run-off from tributaries to the main stream.
- A combined application of the three management options at pinpointed tributaries and river sites may prove to be the best management practice (BMP) in response to urbanization of the Torrens catchment.

The SCS curve number approach performed well in this case study, but the results of streamflow calibration can be improved for densely urbanized sub-catchments by the Green and Ampt method in the SWAT model as suggested by Tasdighi et al. [56]. The results of scenario analyses in this study are restricted by simplified assumptions related to the default configuration of urban land uses in SWAT, and affected by some uncertainty. However, the results are showing most likely trends and magnitudes of expected effects of different land use developments and mitigation solutions on the catchment. Future research will build on outcomes of this study by extending the research to the downstream estuary region in order to address the effects of urbanization, and other potential sources that could combine with urbanization to cause significant threat to the riparian and coastal ecosystems.

Author Contributions: All authors contributed to the study design. Data acquisition and preparation were performed by H.H.N. Development of soil data for modelling was assisted by W.M. Modelling and writing of the original draft was done by H.H.N., F.R. and W.M. reviewed thoroughly the manuscript for submission. The final version of the manuscript was prepared by H.H.N.

Funding: This research was funded by the SA Water Corporation (grant number SW100309) and the Department of Environment, Water and Natural Resources (grant number F0000108271).

Acknowledgments: The authors acknowledge Malcolm Sheard for his professional advice and data support during the development of soil database for this study. We would like to thank Leon van der Linden, Brooke A. Swaffer and Robert Daly for their constructive comments on the study outputs. The authors also wish to thank Margaret Cargill for editing the manuscript for linguistic quality.

Conflicts of Interest: The authors declare no conflict of interest.

References

1. United Nations. *Urban and Rural Areas 2007*; United Nations: New York, NY, USA, 2007. Available online: http://wwwunorg/esa/population/publications/wup2007/2007_urban_rural_chartpdf (accessed on 1 May 2017).
2. Saier, M.H. Are megacities sustainable? *Water Air Soil Pollut.* **2007**, *178*, 1–3. [CrossRef]
3. Xian, G.; Crane, M.; Junshan, S.J. An analysis of urban development and its environmental impact on the Tampa Bay watershed. *J. Environ. Manag.* **2007**, *85*, 965–976. [CrossRef] [PubMed]
4. Evans, C.V.; Fanning, D.S.; Short, J.R. Human-influenced soils. In *Managing Soils in an Urban Environment*; Brown, R.B., Huddleston, J.H., Anderson, J.L., Eds.; ASSA-CSSA-SSSA Publishers: Madison, WI, USA, 2000; pp. 33–67. ISBN 0891181431.
5. Doichinova, V.; Zhiyanski, M.; Hursthouse, A. Impact of urbanization on soil characteristics. *Environ. Chem. Lett.* **2006**, *3*, 160–163. [CrossRef]
6. Whitehead, P.G.; Lapworth, D.J.; Skeffington, R.A.; Wade, A. Excess nitrogen leaching and C/N decline in the Tillingbourne catchment, southern England: INCA process modelling for current and historic time series. *Hydrol. Earth Syst. Sci.* **2002**, *6*, 455–466. [CrossRef]
7. Zhang, Y.; Xia, J.; Shao, Q.; Zhai, X. Water quantity and quality simulation by improved SWAT in highly regulated Huai River Basin of China. *Stoch. Environ. Res. Risk Assess.* **2013**, *27*, 11–27. [CrossRef]
8. Ilman, M.A.; Gell, P.A. *Sediments, Hydrology and Water Quality of the River Torrens: A Modern and Historical Perspectives*; Three Projects for the River Torrens; Water Catchment Management Board; Department of Geographical and Environmental Studies, University of Adelaide: Adelaide, Australia, 1998; pp. 10–53.
9. Clark, T.; Humphrey, G.; Frazer, A.; Sanderson, A.; Maier, H. Effectiveness of flow management in reducing the risk of cyanobacterial blooms in rivers. In Proceedings of the Hydrology and Water Resources Symposium, Melbourne, Australia, 20–23 May 2002; Institution of Engineers: Barton, Australia, 2002.
10. Brookes, J.D. *River Torrens Water Quality Improvement Trial—Summer 2011/12*; Goyder Institute for Water Research Technical Report Series No. 12/4; The Goyder Institute for Water Research: Adelaide, Australia, 2012.
11. Wong, T.H.F. *Australian Runoff Quality—A guide to Water Sensitive Urban Design Engineers Australia*; Engineers Media for Australian Runoff Quality Authorship Team: Queensland, Australia, 2006.
12. Zoppou, C. Review of urban storm water models. *Environ. Model. Softw.* **2001**, *16*, 195–231. [CrossRef]
13. Arnold, J.G.; Srinivason, R.; Muttiah, R.R.; Williams, J.R. Large area hydrologic modeling and assessment Part I: Model development. *J. Am. Water Resour. Assoc.* **1998**, *34*, 73–89. [CrossRef]
14. Gassman, P.W.; Reyes, M.R.; Green, C.H.; Arnold, J.G. The soil and water assessment tool: Historical development, applications, and future research directions. *Trans. Am. Soc. Agric. Eng.* **2007**, *50*, 1211–1250. [CrossRef]
15. Neitsch, S.L.; Arnold, J.G.; Kinir, J.R.; Wiliams, J.R. *Soil and Water Assessment Tool: Theoretical Documentation*; Technical report no 406; Texas Water Resources Institute-Texas A&M University: College Station, TX, USA, 2011.
16. Richards, C.E.; Munster, C.L.; Vietor, D.M.; Arnold, J.G.; White, R. Assessment of a turfgrass sod best management practice on water quality in a suburban watershed. *J. Environ. Manag.* **2008**, *86*, 229–245. [CrossRef]
17. Zhou, F.; Xu, Y.; Chen, Y.; Xu, C.Y.; Gao, Y.; Du, J. Hydrological response to urbanization at different spatio-temporal scales simulated by coupling of CLUE-S and the SWAT model in the Yangtze River Delta region. *J. Hydrol.* **2013**, *485*, 113–125. [CrossRef]
18. Qiu, Z.; Wang, L. Hydrological and Water quality assessment in a Suburban watershed with mixed land uses using the SWAT model. *J. Hydrol. Eng.* **2014**, *19*, 816–828. [CrossRef]
19. Sunde, M.; He, H.S.; Hubbart, J.A.; Scroggins, C. Forecasting streamflow response to increased imperviousness in an urbanizing Midwestern watershed using a coupled modeling approach. *Appl. Geogr.* **2016**, *72*, 14–25. [CrossRef]
20. Schütte, S.; Schulze, R.E. Projected impacts of urbanisation on hydrological resource flows: A case study within the uMngeni Catchment, South Africa. *J. Environ. Manag.* **2017**, *196*, 527–543. [CrossRef]
21. Jordan, Y.C.; Ghulam, A.; Hartling, S. Traits of surface water pollution under climate and land use changes: A remote sensing and hydrological modeling approach. *Earth-Sci. Rev.* **2014**, *128*, 181–196. [CrossRef]

22. Wang, R.; Kalin, L. Combined and synergistic effects of climate change and urbanization on water quality in the Wolf Bay watershed, southern Alabama. *J. Environ. Sci.* **2018**, *64*, 107–121. [CrossRef]

23. Lee, T.; Wang, X.; White, M.; Tuppad, P.; Srinivasan, R.; Narasimhan, B.; Andrews, D. Modeling Water-quality loads to the reservoirs of the Upper Trinity River basin, Texas, USA. *Water* **2015**, *7*, 5689–5704. [CrossRef]

24. El-Khoury, A.; Seidou, O.; Lapen, D.R.; Que, Z.; Mohammadian, M.; Sunohara, M.; Bahram, D. Combined impacts of future climate and land use changes on discharge, nitrogen and phosphorus loads for a Canadian river basin. *J. Environ. Manag.* **2015**, *15*, 76–86. [CrossRef] [PubMed]

25. Saha, P.P.; Zeleke, K. Rainfall-Runoff Modelling for Sustainable Water Resources Management: SWAT Model Review. In *Sustainability of Integrated Water Resources Management Water Governance, Climate and Ecohydrology*; Setegn, S.G., Donoso, M.C., Eds.; Springer International Publishing: Basel, Switzerland, 2015.

26. Vanderkruk, K.; Owen, K.; Grace, M.; Thompson, R. Review of Existing Nutrient, Suspended Solid and Metal Models. Scientific Review for Victorian Department of Sustainability and Environment. Available online: https://ensymdsevicgovau/docs/DSE_EnSym_WQ_Review_RT051009_final_correctionpdf (accessed on 5 January 2017).

27. Shrestha, M.K.; Recknagel, F.; Frizenschaf, J.; Meyer, W. Assessing SWAT models on single and multi-site calibration for the simulation of flow and nutrients loads in the semi-arid Onkaparinga catchment in South Australia. *Agric. Water Manag.* **2016**, *175*, 61–71. [CrossRef]

28. Nguyen, H.H.; Recknagel, F.; Meyer, W.; Frizenschaf, J.; Shrestha, M.K. Modelling the impacts of altered management practices, land use and climate changes on the water quality of the Millbrook catchment—Reservoir system of South Australia. *J. Environ. Manag.* **2017**, *202*, 1–11. [CrossRef]

29. Gale, R.J.B.; Gale, S.J.; Winchester, H.P.M. Inorganic pollution of the sediments of the River Torrens, South Australia. *Environ. Geol.* **2006**, *50*, 62–75. [CrossRef]

30. Shread, M.J.; Borrman, G.M. *Soils, Stratigraphy and Engineering Geology of near Surface Materials of the Adelaide Plains*; Department of Mines and Energy South Australia: Adelaide, Australia, 1996.

31. Australian Soil Resources Information System (ASRIS). Available online: http://www.asris.csiro.au/mapping/viewer.htm (accessed on 1 March 2015).

32. Ritchie, J.T.; Gerakis, A.; Suleiman, A. Simple model to estimate field measured soil water limits. *Trans. ASAE* **1999**, *42*, 1609–1614. [CrossRef]

33. Suleiman, A.; Ritchie, J.T. Estimating saturated hydraulic conductivity from soil porosity. *Trans. ASAE* **2001**, *44*, 235–239. [CrossRef]

34. Meyer, W.S. Land use potential for forestry (Radiata Pine and Blue Gum). In *South East Water Science Review: Prepared for Lower Limestone Coast Taskforce*; South Australian Department of Water, Land and Biodiversity Conservation (DWLBC): Adelaide, Australia, 2010; p. 32.

35. Scientific Information for Land Owners (SILO). Available online: https://www.longpaddock.qld.gov.au/silo/ppd/index.php (accessed on 15 July 2016).

36. AMLR. Available online: http://amlr.waterdata.com.au/Amlr.aspx (accessed on 10 March 2016).

37. Winchell, M.; Srinivasan, R.; Di Luzio, M.; Arnold, J. *ArcSWAT Interface for SWAT 2012, User's Guide*; United States Academic Decathlon (USDA): Mankato, MN, USA, 2013.

38. Arnold, J.G.; Moriasi, D.N.; Gassman, P.W.; Abbaspour, K.C.; White, M.J.; Srinivasan, R.; Santhi, C.; Harmel, R.D.; van Griensven, A.; Van Liew, M.W.; et al. SWAT: Model use, calibration, validation. *Trans. ASABE Am. Soc. Agric. Eng.* **2012**, *55*, 1491–1508. [CrossRef]

39. Abbaspour, K.C.; Johnson, C.; van Genuchten, M.T. Estimating uncertain flow and transport parameters using a sequential uncertainty fitting procedure. *Vadose Zone J.* **2004**, *3*, 1340–1352. [CrossRef]

40. Abbaspour, K.C.; Rouholahnejad, E.; Vaghefi, S.; Srinivasan, R.; Yang, H.; Kløve, B. A continental-scale hydrology and water quality model for Europe: Calibration and uncertainty of a high-resolution large-scale SWAT model. *J. Hydrol.* **2015**, *524*, 733–752. [CrossRef]

41. Ficklin, D.L.; Lue, Y.; Zhang, M. Watershed modelling of hydrology and water quality in the Sacramento River watershed, California. *Hydrol. Process.* **2013**, *27*, 236–250. [CrossRef]

42. Nash, J.E.; Sutcliffe, J.V. River flow forecasting through conceptual models: Part I A discussion of principles. *J. Hydrol.* **1970**, *10*, 282–290. [CrossRef]

43. Beven, K.J. Binley, A.M. The future of distributed models: Model calibration and uncertainty prediction. *Hydrol. Proc.* **1992**, *6*, 279–298. [CrossRef]

44. Moriasi, D.N.; Arnold, J.G.; Van Liew, M.W.; Bingner, R.L.; Harmel, R.D.; Veith, T.L. Model evaluation guidelines for systematic quantification of accuracy in watershed simulations. *Trans. ASABE* **2007**, *50*, 885–900. [CrossRef]

45. Abbaspour, K.C. *SWAT-CUP: SWAT Calibration and Uncertainty Programs—A User Manual*; Swiss Federal Institute of Aquatic Science and Technology: Dübendorf, Switzerland, 2015.

46. DPLG. The 30-Year Plan for Greater Adelaide. A Volume of the South Australian Planning Strategy, Department of Planning and Local Government, 2010. Available online: https://livingadelaidesagovau/__data/assets/pdf_file/0007/278332/The_30-Year_Plan_for_Greater_Adelaide_2010_compressedpdf (accessed on 10 January 2017).

47. Nguyen, H.H.; Recknagel, F.; Meyer, W. Modelling the runoff, nutrient and sediment loadings in the Torrens river catchment, South Australia using SWAT. In Proceedings of the 22nd International Congress on Modelling and Simulation, Tasmania, Australia, 3–8 December 2017; pp. 1–7.

48. Tuppad, P.; Kannan, N.; Srinivasan, R.; Rossi, C.; Arnold, J. Simulation of agricultural management alternatives for watershed protection. *Water Resour. Manag.* **2010**, *24*, 3115–3144. [CrossRef]

49. Moriasi, D.N.; Steiner, J.L.; Arnold, J.G. Sediment Measurement and Transport Modeling: Impact of Riparian and Filter Strip Buffers. *J. Environ. Qual.* **2011**, *40*, 807–814. [CrossRef] [PubMed]

50. Motsinger, J.; Kalita, P.; Bhattarai, R. Analysis of Best Management Practices Implementation on Water Quality Using the Soil and Water Assessment Tool. *Water* **2016**, *8*, 145. [CrossRef]

51. Nielsen, A.; Trolle, D.; Wang, M.; Luo, L.; Han, B.P.; Liu, Z.; Olesen, J.E.; Jeppesen, E. Assessing ways to combat eutrophication in a Chinese drinking water reservoir using SWAT. *Mar. Freshw. Res.* **2013**, *64*, 475–492. [CrossRef]

52. Kasan, N.A. Nutrient Retention Capacity of a Constructed Wetland in the Cox Creek Sub-Catchment of the Mt. Bold Reservoir, South Australia. Ph.D. Thesis, University of Adelaide, Adelaide, Australia, 2011.

53. White, J.D.; Prochnow, S.J.; Filstrup, C.T.; Byars, B.W. A combined watershed-water quality modeling analysis of the Lake Waco reservoir: II. Watershed and reservoir management options and outcomes. *Lake Reserv. Manag.* **2010**, *26*, 159–167. [CrossRef]

54. Fleming, N.; Cox, J.; He, Y.; Thomas, S.; Frizenschaf, J. *Analysis of Constituent Concentrations in the Mount Lofty Ranges for Modelling Purposes*; Technical Report; eWater Cooperative Research Centre: Adelaide, Australia, 2010; pp. 1–37.

55. Brun, S.E.; Band, L.E. Simulating runoff behavior in an urbanizing watershed. *Comput. Environ. Urban Syst.* **2000**, *24*, 5–22. [CrossRef]

56. Tasdighi, A.; Arabi, M.; Harmel, D. A probabilistic appraisal of rainfall-runoff modeling approaches within SWAT in mixed land use watersheds. *J. Hydrol.* **2018**, *564*, 476–489. [CrossRef]

Article

Protecting Coastlines from Flooding in a Changing Climate: A Preliminary Experimental Study to Investigate a Sustainable Approach

Matteo Rubinato [1,2,*], **Jacob Heyworth** [1] **and James Hart** [1,2]

[1]   School of Energy, Construction and Environment, Coventry University, Coventry CV1 5FB, UK; heywort2@cucollege.coventry.ac.uk (J.H.); ac5950@coventry.ac.uk (J.H.)

[2]   Centre for Agroecology, Water and Resilience, Coventry University, Ryton Gardens, Wolston Lane, Coventry CV8 3LG, UK

*   Correspondence: matteo.rubinato@coventry.ac.uk; Tel.: +44-(0)-24-7765-0887

Received: 20 July 2020; Accepted: 28 August 2020; Published: 3 September 2020

Abstract: Rising sea levels are causing more frequent flooding events in coastal areas and generate many issues for coastal communities such as loss of property or damages to infrastructures. To address this issue, this paper reviews measures currently in place and identifies possible control measures that can be implemented to aid preservation of coastlines in the future. Breakwaters present a unique opportunity to proactively address the impact of coastal flooding. However, there is currently a lack of research into combined hard and soft engineering techniques. To address the global need for developing sustainable solutions, three specific breakwater configurations were designed and experimentally compared in the hydraulic laboratory at Coventry University to assess their performance in reducing overtopping and the impact of waves, quantifying the effectiveness of each. The investigation confirmed that stepped configurations work effectively in high amplitudes waves, especially with the presence of a slope angle to aid wave reflection. These results provide a very valuable preliminary investigation into novel sustainable solutions incorporating both artificial and natural based strategies that could be considered by local and national authorities for the planning of future mitigation strategies to defend coastal areas from flooding and erosion.

Keywords: climate change; coastal protection; coastal flooding; sea defence; experimental modelling; sustainability

---

1. Introduction

In the last 140 years, scientific research has established that average sea levels have significantly increased [1–3], and this phenomenon is accelerating. This is a critical issue as even small increases can have devastating effects on coastal habitats [4–7]. Rising sea levels have been identified as a major cause of flooding events across the world [8,9]. Flooding poses a threat to property, safety, and the economic wellbeing of coastal communities [10]. In fact, considering that coastal areas provide a great amount of economic and leisure activities, they contribute significantly to the local and national economy. Thus, more people are continuously attracted to coastal zones contributing to an intense urbanization of these areas. To aggravate this situation, the ecosystems are also threatened by the impact of human activities in coastal areas as well as by the increase of natural extreme weather events (e.g., intensity and duration of storms, floods) generated by climate change, which interfere with local wave climate and changes in morphological beach characteristics [11]. More frequently, high tides reach values that cause costal recession and high sediment transport deficit, and hence, it is necessary to protect these areas with various coastal structures to reduce or at least to mitigate coastal erosion problems. As a result, impacts of climatic variations are usually the greatest along the coast [12–14]. However, many of the current coastal protections (e.g., groins, seawalls, and emerged breakwaters) were built with the single purpose of protecting the coast, without environmental or economic concerns, maintenance costs, or the negative consequences that such structures could cause up to considerable distances along the coast. Coastal regions and their managers consequently face ever-increasing challenges to accommodate safely both the growth of these areas and their development [15].

Traditionally, bulkheads, seawalls, and revetments have been the most commonly used type of shoreline infrastructure implemented as a primary response to coastal hazard. Other applications such as shoreline armouring have also been adopted to protect coastal property from hazards like erosion and flooding [16]. However, there has been a growing interest during the last decade in developing sustainable approaches to guarantee solutions that could deal with the daily and emergency issues in parallel with promoting downtown living [17–19]. For example, in Hong Kong, the land policy emphasizes ecological protection [20–23] and reclamation, enhancing the innovative value in sustainable coastal land use management.

In line with these new approaches, recent studies conducted by scientists and practitioners have demonstrated the benefits of nature-based strategies for restoring degraded coastal ecosystems and mitigating risks including natural defences and "living shorelines" [24,25]. Without any human interaction, shorelines are mainly comprised of biogenic habitats (e.g., saltmarshes, mangroves, oyster and coral reefs) in their natural conditions. These natural coastal habitats secure the provision of essential habitat for marine life, promotion of favourable water quality, and reduction of shoreline erosion and flooding by attenuating waves, stabilizing sediments, and dampening surge [24,26,27]. As such, they are widely valued for their environmental benefits. By adopting alternative sustainable approaches, it is possible to enhance the quality of natural environments along the coasts that can help reduce the impact of coastal hazards [28–32].

It is clear that a crucial goal is to identify nature-based structures that can protect coastal areas and provide a low-cost option to effectively reduce the damaging effects of extreme meteorological events on coastal populations by absorbing storm energy [33], thus enhancing the quality of lives of people living in the surrounding areas. These green areas (including vegetation such as coral reefs or aquatic plants) typical of nature-based solutions could aid the production of sediments (sea grass beds and coral reefs) or could store and hold the sand together (mangroves and coastal dunes) [34]. For example, the benefits provided by coastal herbaceous wetlands in helping to reduce economic damages generated by hurricanes and their impacts have already been demonstrated [34,35].

One type of solution that has not been considered is the mix of artificial and green solutions. Human design structures can guarantee resistance to strong wave impacts and reduce the amount of flooding in coastal areas. However, if mixed with natural ecosystems/green solutions that can still help to reduce wave energy, coastal erosion, and flood hazards [36–41], it could also be possible to recover the natural functioning of the entire coastal area and target future conservation and restoration processes [35–37]. In brief, this option promotes coastal protection through the recovery of the natural functioning of natural ecosystems by means of conservation and restoration actions [38,42]. The trade-offs between socioeconomic development and conservation can be integrated [43–45], which will help with improving coastal development and promoting a sustainable coastal development.

This study provides a comprehensive review of existing hard and soft solutions adopted for coastal protection. Furthermore, it will experimentally investigate and compare preliminary sustainable approaches that could deliver both protection from coastal flooding and the added benefit of conserving, sustaining, and restoring valuable ecosystem functions and services to local communities [46–51].

Hard and Soft Engineering Solutions for Coastal Protection

To identify structural designs that assess new sustainable approaches for coastal protection and to highlight the advantages and disadvantages of existing hard and soft engineering solutions adopted to protect coastal lines, a review was conducted on the techniques available to date. Table 1 summarises the results obtained.

Table 1. A review of existing coastal protection measures with advantages and disadvantages identified for each solution.

Engineering Method	Hard (H) or Soft (S)	Brief Description	Advantages	Disadvantages
Sea wall [52–54]	H	Wall built by the coastline (usually built along the front of cliffs to protect settlements and often curved to reflect wave energy).	• Effectively dissipates wave energy from high impact waves • Long life span and re-assures local communities	• Prevents the movement of beach material along the coast and beach may be lost without replenishment • Maintenance high and expensive to construct
Breakwater [55–57]	H	When waves hit the breakwaters, the power of the wave is dissipated on the breakwater structure so the erosion impact on the cliffs is much less.	• Effectively dissipates wave energy • Easy to maintain	• Prevents the movement of beach material along the coast and beach may be lost without replenishment
Tetrapods [58,59]	H	Multi angular concrete shaped that are preformed and tipped onto the beach to form interlocking components.	• Effectively dissipate energy • Easy installation	• Only applicable at low water level and usually used offshore
Gabions [60–62]	H	Wire cage with pebbles stones and rocks inside. Protect the coastline by reducing the energy of the wave before it directly hits the cliffs	• Allow the build-up of a beach • Easy installation • Relatively cheap to construct • Dissipated wave energy	• Regular maintenance required as faces constant high impact waves • Looks unnatural and not robust
Revetments [62–66]	H	Sloping structures on banks or cliffs built in such a way to absorb some of the energy from the incoming water.	• Effective way of dissipating energy by utilising beach like slope method. • Cheaper and less intrusive than sea walls	• Still allows for erosion to take place • Unsuitable where wave energy is high and difficult to maintain
Groynes [52,67,68]	H	Wooden barrier built at right angles to the beach to retain material and prevent longshore drift.	• Prevents the movement of beach material along the coast (beach encourages tourism) • Relatively cheap to construct	• Unattractive structure • Trapping sediment can prevent the replenishment of sediment further down the coastline increasing erosion elsewhere
Boulder Barrier [69,70]	H	Large boulders piled up on the beach.	• Prevent the effects of coastal erosion effectively • Help to prevent coastal flooding	• Boulders can become easily dislodged with the force of the sea. As a result, they may cause more damage during transportation • Requires regular maintenance

Table 1. *Cont.*

Engineering Method	Hard (H) or Soft (S)	Brief Description	Advantages	Disadvantages
Mangrove Planting [71–74]	S	Mangroves planted along coastline to dissipate wave energy, trap sediment, and control water levels	• Can help to prevent coastal flooding • Can trap pollutants from coming back to land Effective at trapping sediment • Benefits to marine life	• Not effective against high waves. Struggle to adapt to certain climates
Offshore Reefs [75,76]	S	Artificial sand/gravel offshore deposits designed to intercept wave action and dissipate energy.	• Effectively dissipates wave energy • Benefits marine life	• Impact is comparatively a lot less than many hard engineering techniques • Deposits require replacing
Seagrasses [77,78]	S	Submerged aquatic vegetation ecosystem with thick stems.	• Effectively dissipates wave energy • Sustainable solution • Benefits marine life • No maintenance	• Not effective against large storm waves • Seagrasses may be damaged as not protected
Sills [79,80]	S	Shingle or sand beach that is often submerged.	• Effectively dissipates wave energy	• Deposits can often require replacing
Beach nourishment [52,81–83]	S	Replacing beach or cliff material that has been removed by erosion.	• Beaches dissipate wave energy effectively • Easy to monitor impact of longshore drift	• Not sustainable as problem will continue and more material will require replacing • Material
Managed retreat [52,84–86]	S	Allocated areas of the coast that can erode and flood naturally (low value areas)	• Low costs in protection measures	• Loss of land over prolonged period may mean protection measures are required down the line
Beach Dewatering [75]	S	The artificial lowering of the water table within beaches by a system of drains and pumps	• Alternative to more traditional methods of shoreline stabilization • Stabilization of sediments on the surface of the beach • Fast recovery of the beach after storms • Build-up of a sand stock serving as a "buffer-stock" for the following storms	• Expensive • Maintenance • Can contaminate bodies of water

To date, as previously mentioned, natural solutions have been adopted to preserve and/or restore coastal areas. For example, the presence of wetlands has demonstrated to retard waves and the mass flux of water with the presence of vegetation [87]. Despite a few studies on the effect of these vegetated surface, there are not specific guidelines available to determine the optimal shape of the vegetation to consider, the density to be selected, or the height of the vegetation to make it fully under water or emergent. Therefore, to seek this information, this preliminary experimental study was conducted to propose an approach that could combine hard and soft engineering characteristics; thus, it can be the base for a sustainable solution to be adopted. Despite initially using non-real vegetation due to the limitations explained below, hard and soft engineering techniques should be combined in a more ecological way (e.g., facilitating the growth of aquatic plants next to artificial structures), to achieve a less invasive structure on the environment and mitigate the negative influence of hard engineering on ecosystems [49]. In order to identify a feasible "softer" hard sustainable engineered solution, the paper experimentally compared three solutions tested in a wave tank with a physical model, which are presented in Section 2, on the foreshore of the beach and thus did not impede the wave energy or prevented land to sea interaction. The main purpose of the submerged breakwater systems identified is wave attenuation, with the idea of creating splashing and hydraulic conditions that can support sediment capture, helping at the same time in the mitigation of storm surge [30].

2. Materials and Methods

The experimental work presented in this paper was conducted using a wave flume at the Sir John Laing Building, Coventry University (Figure 1). The flume is 18 m long, 1 m deep, and 0.6 m wide. A wave generator is located at the upstream end of the flume while a beach is located at the downstream end to dissipate the energy induced by the waves reproduced.

2.1. Experimental Configurations

To identify sustainable breakwater solutions previously mentioned in Section 1 and investigate their benefits against the use of hard and soft breakwater strategies, three different configurations of sustainable breakwaters (A, B, and C; Figure 2) have been designed and tested within the flume for their effect on overtopping volume and wave attenuation. These sustainable breakwater solutions were tested under a variety of hydraulic wave conditions characterized by dissimilar frequencies and amplitudes.

Figure 1. Wave flume apparatus. Example of wave generation along the flume (**left**), wave generator at the upstream section of the flume (**centre**), and dissipation beach at the downstream section of the flume (**right**).

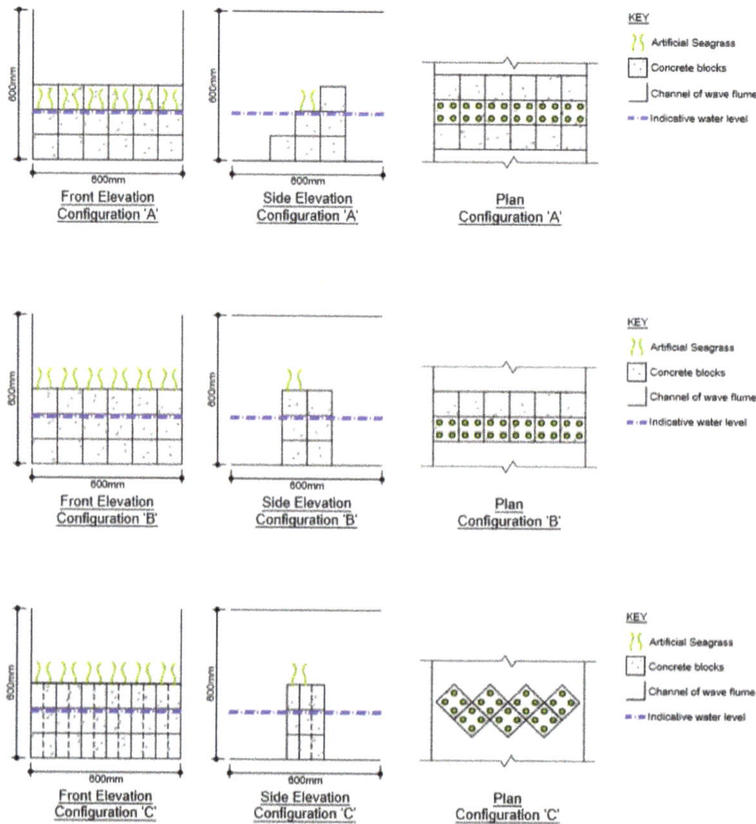

Figure 2. Sustainable coastal protections. Configurations A–C identified in this study.

Configuration A consists of a partly submerged breakwater wall with three steps and artificial vegetation located on the second step of the structure to simulate thick stem vegetation, as displayed in Figure 2. Studies into the wave overtopping of stepped revetments [64] pinpointed that their effectiveness is due to the introduction of slope roughness. Furthermore, it was highlighted that stepped structures, constituting of a slope with uniform roughness, can reduce overtopping volumes of breaking waves up to 60% compared to a smooth slope [64]. This configuration was therefore designed with uniform steps to gradually take the energy out of the wave as the flow could be channelled up the face of the structure. By utilising this approach, the wave collision could be less direct, and water may pass over the structure with less energy rather than generating intense splashing. Vegetation installed on the second step aims to assist with creating increased friction and dissipate wave energy prior to the overtopping. When thinking about reflected waves, the aim is that the sloped shape of the structure could aid destructive interference once the reflected wave meets the incoming waves that they will be out of phase, resulting in the two waves cancelling each other out and giving a reduced wave impact thereafter.

Configuration B is a flat facing and partly submerged breakwater wall with artificial vegetation located on top of the structure (to simulate thick stem vegetation) as shown in Figure 2. This configuration was used to optimize existing hard infrastructures (sea walls) where it would be possible to notice nature adapting to the existing conditions and growing on surfaces not ideal (concrete). Furthermore, this configuration could also replicate the forces interaction between artificial and natural solution where the last layer of the hard structure (seawall) is an ideal environment for coral reefs and porous structures to develop and grow under control. This configuration has been mainly considered to observe which kind of effects could have vegetation on top of existing structures for the simplest case of seawall.

Configuration C is a partly submerged breakwater wall with angled blocks and artificial vegetation located on the top of the structure (to simulate thick stem vegetation), as shown in Figure 2. A study conducted on breakwaters by Ahmadian, 2016 [88], detected several features influencing the effect of the incident wave impact on structures. This work informed that wave breaking, or turbulent losses, can be increased with geometrical alterations, structural characteristics, and the water to structure depth ratio [88]. By incorporating angled blocks, it provided a streamlined method of cutting through incident waves. In turn, this caused waves to become more turbulent, and energy depleted gradually prior to hitting the main body of the wall, rather than causing an instant impact. This configuration allowed comparison of results against the wall shown in Figure 2, to recognise if geometrical alterations, such as streamlining the concrete blocks, assist in dissipating wave energy, in contrast to the high impact stopping force that the flat facing angular wall can offer. Vegetation on the top was intended to dissipate the energy of any overtopping waves.

For each of the three structural configurations displayed in Figure 3, experiments were conducted both with and without a testing platform. The beach in the flume has a gradient of 4.5%. Existing studies expressed [89–91] the importance of a recurved wall profile for high wave return walls, since they define the trajectory of the returned water jet. Shallow angles proved the most effective in attenuating and reflecting waves. Therefore, all the configurations were tested with and without the platform, so that the datasets obtained could have been compared to assess the effectiveness of a slope angle that aims to reflect wave energy.

Figure 3. Overall geometrical configurations.

All the three coastal protection structures tested in this research where built with different configurations of concrete cubes (Figure 3). These had been manufactured from a normal mix with a strength of 20 N/mm^2 (fck) and proportions 1:2:3:0.5, Portland cement, fine aggregate, 10mm coarse aggregate and water. A total of 36 (100 × 100 × 100 mm) cubes were cast and left to cure for 28 days to achieve full strength.

To measure overtopping volumes, a vertical overtopping collection board was manufactured from plywood (600 × 300 × 10 mm), with small arcs at the base, allowing the water to pass freely between either side of the structure. This allowed a detachable metal collection tray (600 × 200 × 100 mm) to be hooked on the plywood wall as demonstrated in Figure 4. The wall was located on the foreshore slope in the flume (14 m) and determined the point at which overtopping was being collected. A ruler (accuracy ±1 mm) was used to measure the height of water in the tray prior to testing and after simulation to allow the change in volume collected to be calculated. From this collection method, a volume was provided in litres for resultant graphs by utilising the following calculation:

$$V_c = (W_w \times L_w \times H_w)/1000 \tag{1}$$

where V_c is the volume collected = overtopping (litres), W_w is the measuring device width (20 cm), L_w is the measuring device length (60 cm), H_w is the measuring device depth measured (cm), and 1000 is the conversion factor used to transform from cubic metres to litres.

As the collection device had a maximum capacity of 12 litres, a measuring jug was used to empty water back into the flume on the side of the incoming wave to ensure the water levels either side of the wall remained constant. The testing platform (600 × 300 mm) for assessing structures with and without a slope angle can also be noticed in Figure 4. This had a varying thickness across its length to account for the sloping foreshore (1 in 20 gradient).

Figure 4. Overtopping collection device. The red box highlights the testing platform.

2.2. Hydraulic Testing Conditions

Two different wave spectrums were used in this study in order to simulate the way different oceans act. This research uses the following wave spectrums within its testing:

- Sine waves simulated regular waves that occur in bodies of water. This aimed to investigate the different structural configurations performed with a regular and repeating low-energy wave. During the tests, frequency and amplitude were varied. To investigate the effect of changing frequency, the frequency ranged from 0.2 Hz up to 0.5 Hz, with overtopping measured at intervals of 20 s. The amplitude was the control variable at 0.05 m. The overall duration of each test was 60 s. The reason for changing the frequency was to assess how each design can influence the reflection of incoming waves to create destructive interference and review its effect on overtopping volumes collected. The experiments then assessed changing amplitudes, where values of amplitude tested ranged from 0.05 m to 0.09 m, in intervals of 0.01 m. As a control measure, the frequency remained at 0.02 Hz throughout (this was the maximum possible due to limitations with the calibration of the equipment tested). Again, the overall testing duration was 60 s. This comprised of a 10 s run time for each experiment, 20 s to allow for the observation of the water, and a further 30 s allowing the water to rest prior to additional testing. The reason for testing change in amplitude was to find patterns to help assess each designs' effectiveness in attenuating and reflecting wave energy under increasing wave height.
- JONSWAP waves to simulate varying waves patterns found in ocean waters, where there are intermittent waves at different frequencies and irregular amplitudes are of a higher energy. This aimed to mimic realistic water effects of varying wave forms on a structure.

By using an off-the-self computer program associated with the control software for the wave tank piston, irregular patterns in waves could be produced in a synthesis to simulate a JONSWAP wave. Table 2, shown below, displays the characteristics of these waves.

Table 2. JONSWAP simulation parameters.

Gamma (γ)	Height of Waves	Amplitude	Period of Waves (Tp)	Max Frequency	Min Frequency
6.6	0.6 m	0.3 m	0.9 s	2 Hz	0.2 Hz

The figures for the JONSWAP synthesis above were chosen to simulate a higher wave energy, compared to that tested in the sine wave experiments. The chosen JONSWAP wave synthesis had a frequency between 0.2 Hz to 2 Hz (compared to 0.2 Hz to 0.5 Hz tested in sine waves) and an amplitude of up to 0.3 m (which is significantly higher than the amplitudes of 0.05–0.09 m tested in the sine waves testing). The purpose of testing in these more extreme conditions was because a JONSWAP simulation relates to irregular wave patterns, where there would likely be a potential storm situation. Table 3 summarises the conditions for all the experimental tests conducted.

Due to the impracticability of growing real seagrasses, a physical model has been made to reproduce submerged vegetation by using straws and plastic sheets to mimic the thick stem structure and broad narrow leaves as shown in Figure 5. Translucent 100 mm straws were used and cut to replicate the 'V' shape for the plastic sheets to slot in. The plastic sheets were fairly stiff and had a course surface providing increased roughness and stood at about 100 mm high making the overall vegetation height 100–150 mm. This was then held together with tape and stuck to the holed board with glue. This kind of flexible setup aimed at representing the binding between interlocking structures that together can create a more sustainable barrier needed to combat the wave energy towards the beach to be protected, as well as miming the behaviours of reefs and submerged vegetation. However, it is also essential to consider the limitations associated with the choice of not using actual seagrass. By using similar structures next to each other, realistic and complex plant morphologies such as flexing elements with varying cross-sectional area over depth could not be replicated, leading to dissimilar flow patterns generated by a variety of stems, branches, roots, and leaves. Even if the height of the stems or the length of the roots can interfere with erosion, deposition patterns, transport of pollutants, stability of the plant, and exchange of nutrients between one type of vegetation to another, this was not the main focus of the study presented in this paper.

The choice of this artificial solution was made to isolate specific responses within the laboratory experiment under controlled conditions and to inform future work with real vegetation. Ideally, future studies will also incorporate the testing of specific patches and geometries which could generate a variety of drag coefficients C_D and Reynolds numbers Re.

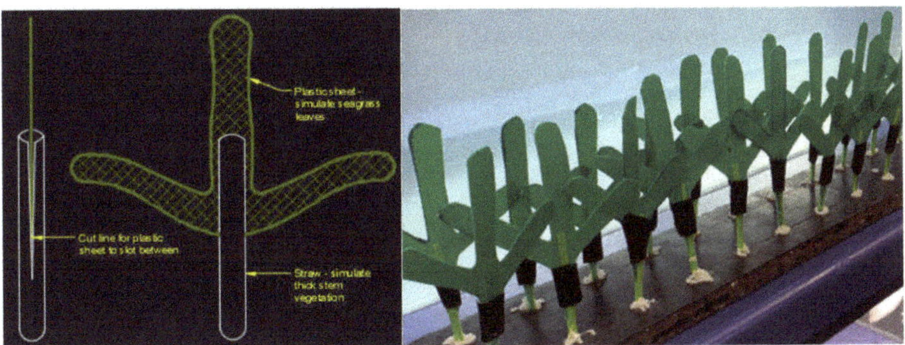

Figure 5. Artificial sea grass reproduced.

Testing was repeated three times for each hydraulic condition and corresponding structural configuration simulated. Simulations were also recorded using a camera to allow further analysis of the hydraulic behaviours (e.g., wave impact on the protective structures).

Table 3. Experimental testing conditions.

Analysis	Hydraulic Conditions	Structural Configuration	Testing Platform Used
Frequency Vs Overtopping	Sine Spectrum	A	Yes
Frequency Vs Overtopping	Sine Spectrum	B	Yes
Frequency Vs Overtopping	Sine Spectrum	C	Yes
Frequency Vs Overtopping	Sine Spectrum	A	No
Frequency Vs Overtopping	Sine Spectrum	B	No
Frequency Vs Overtopping	Sine Spectrum	C	No
Amplitude Vs Overtopping	Sine Spectrum	A	Yes
Amplitude Vs Overtopping	Sine Spectrum	B	Yes
Amplitude Vs Overtopping	Sine Spectrum	C	Yes
Amplitude Vs Overtopping	Sine Spectrum	A	No
Amplitude Vs Overtopping	Sine Spectrum	B	No
Amplitude Vs Overtopping	Sine Spectrum	C	No
Overtopping Vs Time	JONSWAP Spectrum	A	Yes
Overtopping Vs Time	JONSWAP Spectrum	B	Yes
Overtopping Vs Time	JONSWAP Spectrum	C	Yes
Overtopping Vs Time	JONSWAP Spectrum	A	No
Overtopping Vs Time	JONSWAP Spectrum	B	No
Overtopping Vs Time	JONSWAP Spectrum	C	No

3. Results

This section presents a description of the experimental results, their interpretation, as well as the experimental conclusions that can be drawn.

3.1. Sine Wave Conditions—Frequency Analysis

Resultant data from the testing of overtopping against change in frequency are displayed in Figure 6 (no slope angle) and Figure 7 (with slope angle) below.

To identify a process which could directly provide a comparison between the performances of each structure tested, for each set of frequencies run within the experimental facility, these values have been normalized by using the maximum frequency used, which corresponds to 0.5 Hz. The same procedure was conducted for the overtopping values, which were normalized by using the maximum overtopping amount recorded within the entire set of tests under each configuration. Table 4 displays the experimental datasets collected for these hydraulic conditions.

From the data presented in and Figures 6 and 7, it can be seen that all data sets show an initial increase in overtopping with wave frequency, which obtains a peak value and then decreases with wave frequency. A polynomial second order trend line has been fitted to the data to demonstrate this trend. For tests with no slope angle, Configuration A first obtains the peak value, followed by C and then B. For tests with a slope angle, the peak of Configuration C shifts notably, meaning that now Configuration C is the first to hit peak value, followed by A and then B.

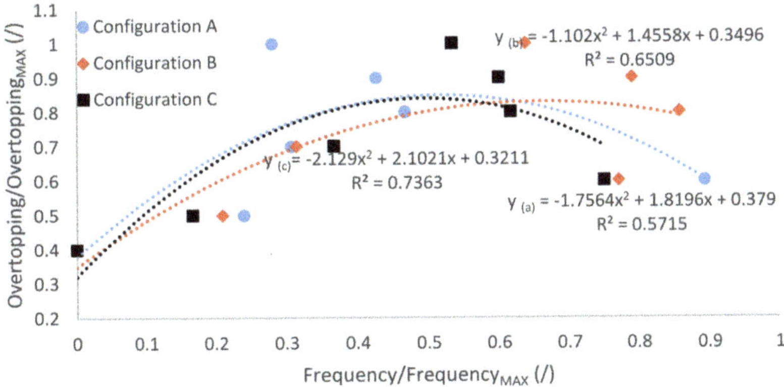

Figure 6. Sine wave hydraulic conditions; relationship between wave crest amplitude A and overtopping volume Q (averaged results); no slope angle adopted within the experimental facility.

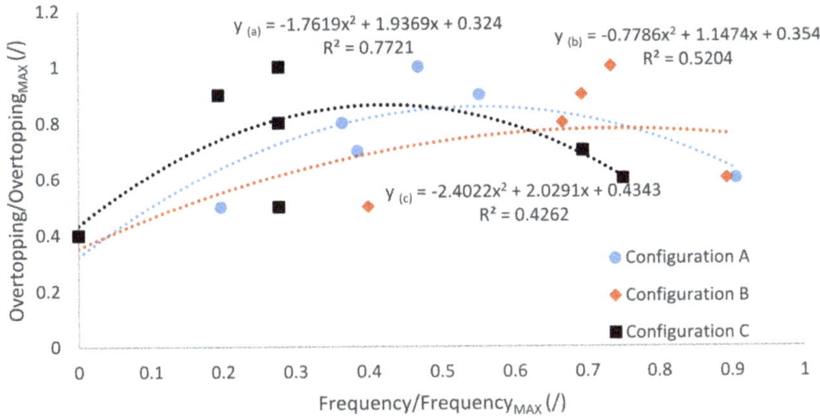

Figure 7. Sine wave hydraulic conditions; relationship between wave crest amplitude A and overtopping volume Q (averaged results); slope angle adopted within the experimental facility.

Table 4. Experimental testing parameters collected for sine wave (F = frequency) with and without slope angle.

F	Conf. A	Conf. B	Conf. C	Conf. A	Conf. B	Conf. C
(Hz)	Overtopping Volume (L)	Overtopping Volume (L)	Overtopping Volume (L)	Overtopping Volume (L)	Overtopping Volume (L)	Overtopping Volume (L)
	No slope angle			With slope angle		
0.2	0	0	0	0	0	0
0.2	0	0	0	0	0	0
0.2	0	0	0	0	0	0
0.25	0.6	0.6	0.36	0.6	1.2	0.24
0.25	0.6	1.2	0.24	0.96	1.44	0.6
0.25	0.96	0.84	0.6	0.72	0.96	0.36
0.3	3	3	1.8	3	2.4	0.6
0.3	2.4	3.6	1.2	3.6	3	1.2
0.3	2.64	3.12	2.4	3.84	2.64	1.44
0.35	0.6	1.2	0.6	1.2	1.8	0.6
0.35	1.2	1.56	1.2	1.44	2.4	1.2
0.35	0.96	1.2	0.84	1.8	2.04	1.2

Table 4. Cont.

F	Conf. A	Conf. B	Conf. C	Conf. A	Conf. B	Conf. C
(Hz)	Overtopping Volume (L)	Overtopping Volume (L)	Overtopping Volume (L)	Overtopping Volume (L)	Overtopping Volume (L)	Overtopping Volume (L)
	No slope angle			With slope angle		
0.4	1.2	3.6	1.2	1.2	1.8	0.24
0.4	1.2	4.2	1.8	1.8	2.4	0.6
0.4	1.8	3	1.44	1.2	1.8	0.36
0.45	1.2	3.6	1.2	2.4	2.4	0.24
0.45	1.2	3	1.44	1.8	1.8	0.36
0.45	1.44	3.36	1.68	2.16	2.04	0.24
0.5	0.6	2.4	1.2	1.8	1.8	0.36
0.5	1.2	3	1.44	1.56	2.4	0.24
0.5	0.72	2.64	1.2	2.04	2.4	0.6

3.2. Sine Wave Conditions—Amplitude Analysis

As shown in Figures 8 and 9 (results summarised in Table 5), Configuration C was the most effective at attenuating wave energy and has the least overtopping volume, closely followed by Configuration B.

Configuration A was the least effective at attenuating wave energy, as the overtopping volumes measured greatly exceeded that of the other configurations, often with the overtopping device reaching full capacity in large amplitude waves.

All configurations showed a linear increese in overtopping with wave amplitude.

Regression analyses presented in Figures 8 and 9 all show correlation values of $R^2 > 0.93$. There is only a slight change in results when a slope angle is present that becomes increasingly evident under large amplitudes exceeding 0.07 m. This indicates that when the structures are subject to high amplitude waves, the effect of a slope angle is more important as the resultant wave shape can be reflected back away from the structure rather than in a vertical profile.

High amplitude waves also have increased energy, so the importance of reflecting this wave energy is emphasised.

Table 5. Experimental testing parameters collected for sine wave (A = amplitude) with and without slope angle.

A	Conf. A	Conf. B	Conf. C	Conf. A	Conf. B	Conf. C
(m)	Overtopping Volume (L)	Overtopping Volume (L)	Overtopping Volume (L)	Overtopping Volume (L)	Overtopping Volume (L)	Overtopping Volume (L)
	No slope angle			With slope angle		
0.05	0	0	0	0	0	0
0.05	0	0	0	0	0	0
0.05	0	0	0	0	0	0
0.06	0.6	1.2	0.36	0.6	1.2	0.24
0.06	0.84	1.44	0.36	0.6	1.56	0.6
0.06	0.6	0.84	0.6	0.84	0.96	0.36
0.07	4.2	2.4	1.2	3.6	3	1.2
0.07	4.8	3	1.44	4.2	3.24	1.56
0.07	4.2	2.64	1.2	4.56	3.6	1.56
0.08	7.2	3.6	1.8	7.2	3.6	1.8
0.08	7.8	3.84	2.4	6.6	3.84	2.4
0.08	8.4	4.2	2.4	7.56	4.2	1.8
0.09	12	4.2	3	12	4.2	2.4
0.09	12	4.8	2.4	10.8	4.8	3
0.09	12	4.56	3	9.6	4.8	2.88

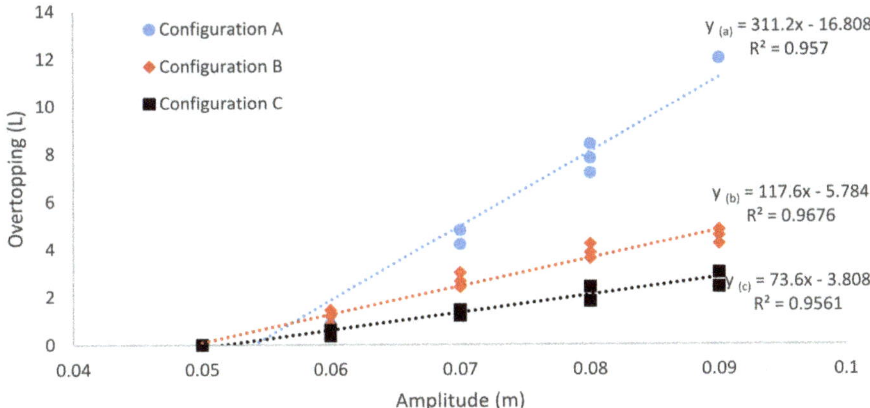

Figure 8. Sine wave hydraulic conditions; overtopping measure vs. amplitude; no slope angle adopted within the experimental facility.

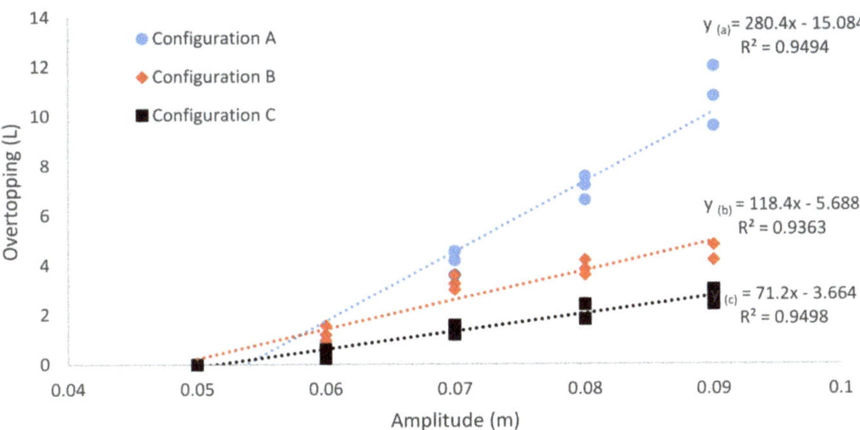

Figure 9. Sine wave hydraulic conditions; overtopping measure vs. amplitude; slope angle adopted within the experimental facility.

3.3. JONSWAP Wave Conditions

Figures 10 and 11 display the comparison of experimental datasets collected under JONSWAP hydraulic conditions without and with a slope angle present (measurements are summarised in Table 6).

Results show that Configuration A was the most effective at attenuating wave energy and had the least overtopping volume collected, closely followed by configuration C. Configuration B was the least effective as overtopping measured greatly exceeded that of the other configurations, with it being unable to complete the full simulation without a slope angle present due to the overtopping device being at full capacity at three minutes (180 seconds) in. It is interesting to note that configurations B and C effectively switch places between tests with the sine wave and JONSWAP wave.

A reduction in overtopping volumes of configurations B and C is noticed when a slope angle is present. Configuration A shows a slight increse in overtopping volume when the slope angle is present.

A linear trendline had been used for graphical data to show a direct correlation between the increase in time and overtopping, and R^2 values obtained exceed 0.91 and are a strong indicator of direct proportionality, despite varying wave heights and frequencies.

The resultant graphs for the JONSWAP simulation against Configurations A, B, and C reinforce the findings from testing in frequency and amplitude. Configuration A and C did not benefit from having a slope angle present, but Configuration B did, as the nature of its shape allowed the reflected wave to be directed away from the face of the structure. This is also noticeable in Figure 10 where it is clear that Configuration B without any slope angle could not complete the full final simulation. Results recorded after the collection tray had reached full capacity have been omitted from the graphical data to give a more accurate trendline as the data was clearly outlying in Figure 11.

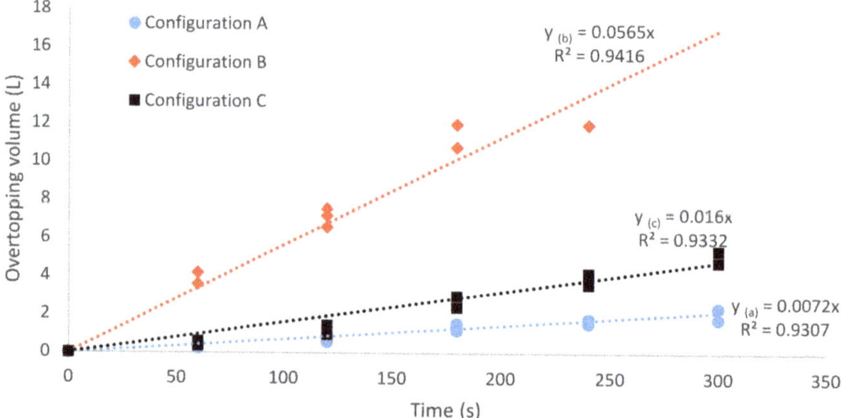

Figure 10. JONSWAP hydraulic conditions, overtopping measure; no slope angle adopted within the experimental facility.

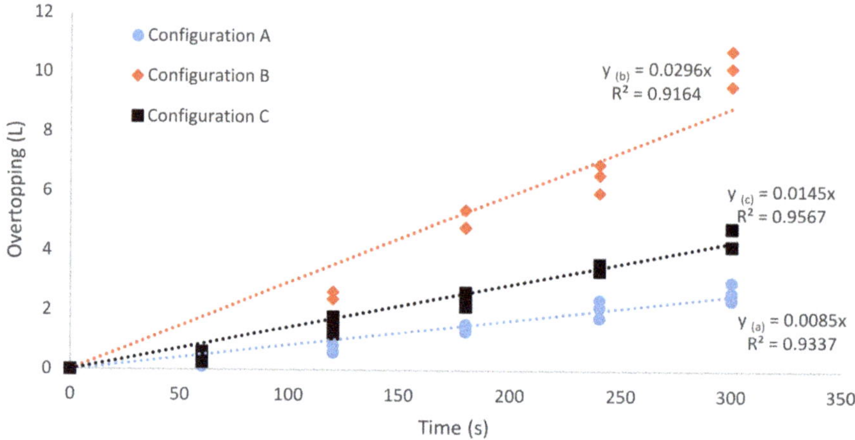

Figure 11. JONSWAP hydraulic conditions; overtopping measure; slope angle adopted within the experimental facility.

All these aspects can be clearly noticed in Figures 12–14 where the performace of each configuration is compared with and without slope angle.

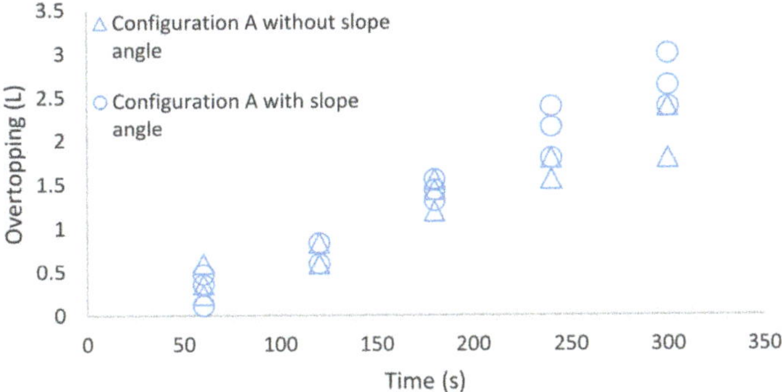

Figure 12. Performance of Configuration A with and without slope angle for JONSWAP hydraulic conditions.

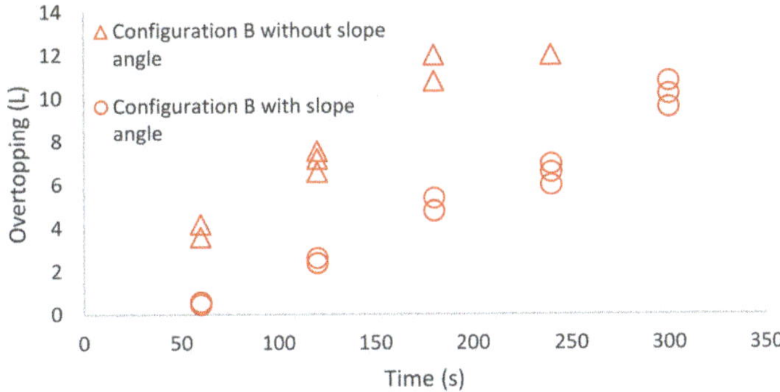

Figure 13. Performance of Configuration B with and without slope angle for JONSWAP hydraulic conditions.

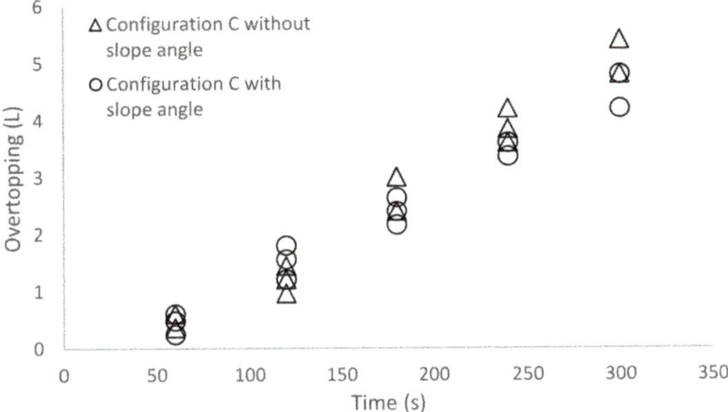

Figure 14. Performance of Configuration C with and without slope angle for JONSWAP hydraulic conditions.

Table 6. Experimental testing parameters collected for JONSWAP waves with and without slope angle.

	Conf. A	Conf. B	Conf. C	Conf. A	Conf. B	Conf. C
Time (s)	Overtopping Volume (L)	Overtopping Volume (L)	Overtopping Volume (L)	Overtopping Volume (L)	Overtopping Volume (L)	Overtopping Volume (L)
	No slope angle			With slope angle		
60	0.24	3.6	0.6	0.12	0.48	0.48
60	0.6	4.2	0.6	0.36	0.6	0.6
60	0.36	3.6	0.36	0.48	0.6	0.24
120	0.6	7.2	1.2	0.6	2.4	1.56
120	0.84	6.6	0.96	0.84	2.4	1.2
120	0.6	7.56	1.44	0.84	2.64	1.8
180	1.2	10.8	2.4	1.32	4.8	2.4
180	1.56	12	3	1.44	5.4	2.16
180	1.44	10.8	3	1.56	4.8	2.64
240	1.56	12	3.6	1.8	6	3.6
240	1.8	/	3.84	2.16	6.6	3.36
240	1.8	12	4.2	2.4	6.96	3.6
300	1.8	/	4.8	2.4	9.6	4.2
300	2.4	/	5.4	2.64	10.8	4.8
300	2.4	/	5.4	3	10.2	4.8

4. Discussion

4.1. Wave Attenuation Mechanisms Observed

Figure 15 displays images taken from lab recordings of high amplitude waves observed during testing. By observing the wave interaction with the structure, it can help us understand why different shaped structures work better in dissipating wave energy and re-directing the incoming water.

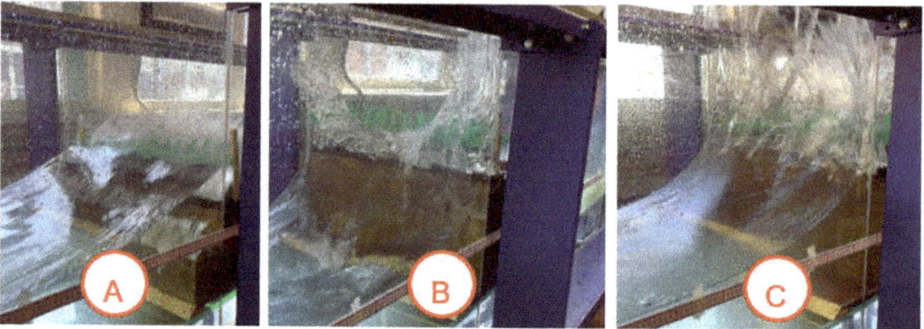

Figure 15. Resultant wave shapes for Configurations A–C.

The behaviours of these waves can be described as follows:

Configuration A—Wave impact was low and flat, resulting in wave energy being dissipated on the breakwater structure. The stepped approach acted as a ramp channelling the water over the top of the structure. However air voids between steps helped to increase turbulence and reduce wave energy. The photographs demonstrate that the artificial vegetation reduces the energy of waves as the stems and broad leaves could be seen to bent back in. This supported Kerpen's claims [64] that stepped structures, constitutive of a slope with uniform roughness, reduce overtopping volumes [64]. Waves were not observed at a great height over the structure and never neared the top of the flume walls.

Configuration B—Wave impact on this structure was sudden and as a result caused the waves to ride up the surface of the flat faced wall. This meant that reflected waves often passed over the structure or collapsed on top in a large wave wall without the presence of a slope angle to direct flow away. The wave height observed was far greater than the other configurations in particular with the configuration tested with 0.8 m amplitude.

Configuration C—The impact of waves was sudden and often had a clapping noise as it impacted the angled block wall and water filled the air voids. The incident wave ran up the surface of the structure and fell in streaks due to the "V" channels created by streamlining the blocks. The wave height observed for a 0.8 m amplitude wave was high, splashing above the flume walls (0.6 m).

Effect of Slope Angles

Having a slope angle was key to real life schemes as often sea defence structures are built on the foreshore and the topographical levels on the ground have varying gradients. At some point in the construction process there will be a decision made whether a platform (structural foundation) is required due to ground conditions and the most suitable angle to aid the protection of the coast and provide stability. From lab testing the key benefits of the shallow slope angle can be summarised as follows.

Surface runoff is directed back out to sea. Potential water that would have overtopped the structure due to surface runoff was directed back towards the incoming waves. Although ultimately this did not make a significant difference to the volume collected within this study, this is important when considering a scaled-up model. Over a longer duration, a large amount of water has the potential to be accumulated, giving an increased importance to last resort defence features, such as sea walls.

Wave reflection is aided and splash is directed back to sea. Rather than the wave splash being at 90° to the water surface and a horizontal splash profile that causes much of the wave to collapse back onto the structure, the introduction of a slope angle means that the resulting splash will be at an acute angle to the water's surface. The wave energy therefore will be directed back out towards incoming waves. The effect of this can be appreciated in the results from the JONSWAP synthesis analysis that with a slope angle Configuration B performed far better, completing a full five-minute simulation that it was not previously able to.

4.2. Effectiveness in Reducing the Overtopping

In order to evaluate the overall effectiveness of structures and assess how they performed in wave attenuation across the various testing spectra, Table 7 was created. It displays a point scoring system based on the overtopping volume collected in resultant graphs, with structures collecting the least water volume being 1st (3 points), 2nd (2 points) and the structure overtopping the most receiving 3rd (1 point).

The total effectiveness in this study concludes that Configuration C performed the best across the three testing scenarios but does not necessarily mean that it is the most practical to use in every coastal scenario. This is due to effectiveness being dependant on multiple conditions including the type of waves the structures are subject to, the location of the protection measure, and subsequent impacts to the ecosystem from its construction.

Table 7. Effectiveness scoring.

Configuration	Frequency Testing		Amplitude Testing		JONSWAP Testing		Points Total	Total
	Slope Angle	No Slope Angle	Slope Angle	No Slope Angle	Slope Angle	No Slope Angle	(/)	%
A	2	1	1	1	3	3	11	30.56
B	1	2	2	2	1	1	9	25.00
C	3	3	3	3	2	2	16	44.44

After considering the results from the sine testing (changing amplitude and frequency), it would have been reasonable to predict that Configuration C would have also been the most effective in a JONSWAP testing scenario. However, this was not the case in JONSWAP testing where Configuration

A outperformed all structures when subject to high energy wave conditions at irregular amplitudes and frequencies. This was mainly because the sine testing was more influenced by friction and gravity (than wave reflection) as the lower energy of the waves had a smaller impact in this respect. In contrast to this, JONSWAP waves simulated high energy waves, which created more interference with each other over the duration. Although friction factors and gravity losses still played a significant part in the JONSWAP simulation, the way the structures reflected wave energy and the resultant wave interception were more important when analysing the performance of configurations tested.

When reviewing the footage of the experiments, interference caused by the reflected wave played a big part in its effectiveness as it created wave interference when two waves from opposite directions meet. When considering Configuration A, the most effective in JONSWAP testing, it could be seen that reflected waves caused destructive interference. The crest of the reflected wave lined up with the trough of the incoming wave, resulting in them cancelling out as they were out of phase and thus creating a reduced wave. On camera footage, the sloped shape of this configuration allowed some overtopping but also allowed some of the incident wave energy to run back down the structure. As a result, this created a rocking motion within the water and aiding the waves sinusoidal wave movement. Another observation during JONSWAP testing is how the reflected wave location moved position in the tank. At the start of testing, the location of reflected waves meeting incoming waves was near to the structure, and as the frequent waves continued, the reflected wave moved back throughout the flume. This indicates that by using structures that are effective at creating destructive interference (Configuration A), the impact on the coastal structure will be lessened and over time overtopping will be greatly reduced as a result of this.

This contrasted to Configuration's B and C, which were not as effective in this process. Due to the nature of their shape creating a high impact force for waves, the wave reflection was more aggressive, unlike the stepped shape breaking down energy and creating turbulence, as the water energy is re-directed up in the air and crashes down. This would often cause constructive interference, making irregular larger waves as a result of the crests of reflected waves and incoming waves lining up. This would help to explain why wall-like structures (such as Configurations B and C) are more effective as a last resort defence on the shoreline, rather than a breakwater on the foreshore. In amplitude testing, R^2 values were taken very close to 1 (direct proportionality). This indicated a very good positive correlation in results, indicating that with increased amplitudes, the wave speed and energy increases, causing a higher overtopping. Configurations with a large impact stopping force, such as B and C, performed far better in these scenarios as they reflected wave energy effectively.

4.3. Real Life Implications

When comparing configuration A to existing structures identified by the literature review, it is possible to see similarities to a coastal revetment. The stepped nature of the structure made it act like a ramp, aiding in dissipating some of the wave energy and proving a direction for the water to travel, so the water runs up its surface, rather than producing a direct impact, by utilising a sloped approach method. Similarities can also be drawn with the tetrapod's strategy as the nature of waves breaking against the structure aiding its wave attenuation can be drawn, and both structures seem most applicable at low water levels, as the stepped structure did not perform well under high amplitude waves.

On the other hand, Configuration B, if compared to existing structures identified by the literature review, has multiple similarities to a seawall structure. It proved effective against high amplitude waves as it provided a direct stopping force for the energy. For real life implications, seawalls are usually curved at the top as the large wave wall produced can then be directed back out to sea. Instead, artificial seagrass located on top of the wall aimed to re-direct water back away from the structure. This method was effective under low wave amplitudes; however, as the wave energy increased, the water was overpowering and often bypassed the seagrass completely due to the reflected water trajectory. It was noticed that the nature of a seawall is not effective in creating destructive interference as when reflected waves met incoming waves; this often led to the creation of larger waves, with higher wave energy and the potential to cause more erosion.

Finally, when comparing configuration C to existing structures identified by the literature review, you can see similarities to a seawall and a breakwater. It could effectively manage high amplitude waves as it could take the high impact of the waves and channel the water up the wall like a seawall. As with Configuration B, the artificial seagrasses located on top appeared to be most effective under low wave energy, where splash height was low and overtopping less aggressive. It also acted in a similar way to breakwater, as the concrete blocks in breakwater are often in random arrangements causing the water to interact with the edges of blocks causing a streamlined effect and channelling the water round them rather than a direct impact with their flat face. This causes the wave energy to disperse rather than a direct impact.

When investigating the sustainability of all the configurations tested, they can be deliberately considered to manipulate the shoreline to satisfy human need [92] and so are still largely seen as hard from an engineering perspective. However, they can all be considered ideal for the development of coral reefs and natural ecosystems that could replace the "green areas" simulated on this study, in line with the theory of incorporating natural habitats into hard solutions by permitting space for coastal adjustments. By implementing sea life and habitat restoration on the foreshore of beaches to combine with engineering options, a combined solution can be found where the ecosystem and engineering methods can act together to provide effective wave attenuation [93,94].

4.4. Limitations

4.4.1. Importance of Slope Factors

The slope of the coast is a key factor that could largely influence the inundation during a flooding event (permanent or sporadic) generated by sea level rise. Additionally, the angle of the beaches could actually control the velocity with which the sea withdraw in case of inland water running for flooding due to other types (e.g., river or urban). This is a crucial factor that was not considered in this study but that will require an extensive experimental campain to produce map of slopes and the consequent hydraulics conditions associated for varius flow rates and velocities to be used to calibrate and validate numerical models and to identify solutions, which could reduce the vulnerability of lower slopes (in the case of flooding from the sea) or higher slopes (in the case of inland flooding) [95,96]. Furthermore, to accurately quantify wave energy and other crucial parameters, more sophisticated equipment is needed. For example, for quantifying the wave energy, an instrument more accurate than a ruler would be necessary to estimate the significant wave height. Low-cost techniques recently published and applied to other fields [97–100] will provide a support in improving the accuracy of the measurement within this study. For example, by using low cost cameras (GoPro), it will be possible to implementing Particle Image Velocimetry and Planar Concentration Analysis techniques to better quantify velocity field and pollutant maps to assess the performance of coastal structures in terms of wave attenuation and pollutant transport.

4.4.2. Importance of Permeability Factors

Studies conducted to date have confirmed that tsunamis and storms have generated washover deposits across beaches or dunes in the last decade [101]. The deposition of sediments therefore continues to alter the morphology of coastal areas after each storm event [102–108], penetrating into existing material and causing various levels of stratification which vary the permeability of the site. This is another aspect that was beyond the scope of this study but would require the characterization of sedimentary characteristics of varius type of washover successions for multiple coastal tophography configurations, including the beach ridge elevation and backshore tophography. The presence of specific permeable material within the first layers of the stratification could in face, if characterized, be used as a sustainable solution for storing part of the water that inundates communities living in coastal areas.

4.4.3. Importance of Marine Currents and Bathymeric Factors

Wind waves, storm surges and ocean circumation play a significan contribution to to risk of flooding in coastal areas [109]. All these aspects can alter the mechanical force of the storm surge [110–113], generating different erosion effects and flooding conditions [114,115]. Despite being typical and dissimilar for each site conditions, concurrence of astronomical high tides and energetic waves can influence the likelihood of overtopping and consequent inundation, posing a hugh threat for coastal population and urbanisation. This aspect requires the quantification of velocity vector maps, quantification of tide rise and the characterization of waves induces by strong winds, and this was not possible to replicate within the experimental facility adopted in this study. However, it is also vital to estimate the interaction between these natural and environmental conditions and the frequency and magnitude of flooding events to target specific schemes that could better perform and are less sensitive to the natural processes involved and their interaction [116].

4.4.4. Importance of Real Vegetation Studies

As previously written, due to the impracticability of growing real seagrasses, a physical model has been made to reproduce submerged vegetation by using straws and plastic sheets to mimic the thick stem structure and broad narrow leaves. The choice of this artificial solution was made to isolate specific responses within the laboratory experiment under controlled conditions and to inform future work with real vegetation. Ideally, future studies will also incorporate the testing of specific patches and geometries, which could generate a variety of drag coefficients C_D and Reynolds numbers Re.

5. Conclusions

The purpose of the research was to assess the viability of a combined hard and soft engineered breakwater solution for coastline protection. A comprehensive literature review was conducted to identify existing structures to aid the protection of coastlines and innovative solutions being investigated worldwide. Advantages and disadvantages for each solution were discussed and combined into three newly designed configurations. Experimental tests were then conducted testing these three different configurations for overtopping performance against a range of varying wave simulations that were designed to replicate different real-life conditions.

The tests were performed at the same testing location, with overtopping measured at the end of each wave simulation to judge the amount of wave attenuation of each structural configuration. The results showed that configurations with a high impact stopping force (such as Configurations B and C) outperformed a stepped structure (Configuration A) in lower energy sine waves that simulate shallower water. During the JONSWAP simulation, however (with higher energy waves, such as would be found in conditions in the North Sea), a stepped configuration outperformed the walled configurations as it attenuated the waves further and hence allowed less overtopping. It was identified that the contributing factor influencing the increased effectiveness was the structure's ability to reflect waves in a nature that causes destructive interference of the reflected wave and the incident wave. This resulted in reduced waves as they cancelled each other out.

In addition to measuring overtopping volumes, a video camera was used to observe the hydraulic behaviours for each structural configuration. These could best be seen under the high amplitude (0.09 m) sine spectrum waves tested, where the increased wave height resulted in increased wave energy. Images provided demonstrate the resultant wave shape of the stepped configuration was low and flat, making it suitable as a breakwater; however, wave impact on a flat faced wall was sudden and caused the waves to ride up the surface. To build further on this, the experiments also explored the performance of each structural configuration with and without using a testing platform. This modification was incorporated to create an angle to the structure in the water, to match that of the sloping foreshore. It was found that the presence matching the sloping foreshore (4.5% gradient) aided structural protection measures with a high impact stopping force (Configurations B and C), with key benefits to the reflected wave trajectory and surface runoff. The findings of this work helped provide recommendations for future research needed to achieve sustainable approaches in coastal defence design.

Future research could explore the performance of the breakwater structures in the remaining ranges of the JONSWAP wave that were not covered in the initial sine testing (by testing frequencies between 0.5–2 Hz and amplitudes from 0.1–0.3 m), in order to better understand and predict the exact frequency and amplitude values, at which the stepped breakwater began to outperform the wall-like structures. Furthermore, in order to further understand sustainable design of submerged breakwaters, future research should focus on the following criteria to be analysed:

- The use of different materials to identify how material roughness influences overtopping and if a sustainable material can be utilized for practical implications.
- The use of real vegetation to investigate effects of flexible coral reefs and underwater vegetation for the wave attenuation and the spread of pollutants in the proximity of coastal areas.
- The testing of structural configurations with different vegetation appropriate for saltwater to assess their effectiveness in reducing overtopping, decreasing wave energy and the structure's effect on their longevity.
- Further experimentation with slope angles to determine a best shape/angle to reflect wave energy with each breakwater design.
- Investigation into sediment movements by testing structures with a hit and miss concrete base.

By allowing these open channels within the structure, the flow of water will work with the natural movement of sands and waves to allow sand deposition further along the coast. This way, the sea defence will not prevent the beach from replenishing its supply of sand as a natural defence to dissipate wave energy. This method will also allow the possibility to investigate longshore drift and the effect of the structure on the movement of beach sediment.

Author Contributions: Conceptualization, M.R.; methodology, M.R. and J.H. (Jacob Heyworth); validation, M.R., J.H. (Jacob Heyworth), and J.H. (James Hart); formal analysis, J.H. (Jacob Heyworth) and M.R.; investigation, J.H. (Jacob Heyworth) and M.R.; resources, M.R.; data curation, M.R., J.H. (Jacob Heyworth), and J.H. (James Hart); writing—original draft preparation, M.R., J.H. (Jacob Heyworth), and J.H. (James Hart); writing—review and editing, M.R., J.H. (Jacob Heyworth), and J.H. (James Hart); visualization, J.H. (Jacob Heyworth); supervision, M.R.; project administration, M.R. All authors have read and agreed to the published version of the manuscript.

Funding: This research received no external funding.

Acknowledgments: The authors would like to thank Ian Breakwell and Craig Harrison for technical support with this project.

Conflicts of Interest: The authors declare no conflict of interest.

References

1. Scardino, G.; Sabatier, F.; Scicchitano, G.; Piscitelli, A.; Milella, M.; Vecchio, A.; Anzidei, M.; Mastronuzzi, G. Sea-Level Rise and Shoreline Changes Along an Open Sandy Coast: Case Study of Gulf of Taranto, Italy. *Water* **2020**, *12*, 1414. [CrossRef]
2. Martínez-Graña, A.; Gómez, D.; Santos-Francés, F.; Bardají, T.; Goy, J.L.; Zazo, C. Analysis of Flood Risk Due to Sea Level Rise in the Menor Sea (Murcia, Spain). *Sustainability* **2018**, *10*, 780. [CrossRef]
3. Chen, W.-B.; Chen, H.; Lin, L.-Y.; Yu, Y.-C. Tidal Current Power Resources and Influence of Sea-Level Rise in the Coastal Waters of Kinmen Island, Taiwan. *Energies* **2017**, *10*, 652. [CrossRef]
4. Melo de Almeida, L.P.; Almar, R.; Meyssignac, B.; Viet, N.T. Contributions to Coastal Flooding Events in Southeast of Vietnam and their link with Global Mean Sea Level Rise. *Geosciences* **2018**, *8*, 437. [CrossRef]
5. White, E.D.; Meselhe, E.; Reed, D.; Renfro, A.; Snider, N.P.; Wang, Y. Mitigating the Effects of Sea-Level Rise on Estuaries of the Mississippi Delta Plain Using River Diversions. *Water* **2019**, *11*, 2028. [CrossRef]

6.   Van De Lageweg, W.I.; Slangen, A.B.A. Predicting Dynamic Coastal Delta Change in Response to Sea-Level Rise. *J. Mar. Sci. Eng.* **2017**, *5*, 24. [CrossRef]

7.   Davtalab, R.; Mirchi, A.; Harris, R.J.; Troilo, M.X.; Madani, K. Sea Level Rise Effect on Groundwater Rise and Stormwater Retention Pond Reliability. *Water* **2020**, *12*, 1129. [CrossRef]

8.   Kumbier, K.; Carvalho, R.C.; Woodroffe, C.D. Modelling Hydrodynamic Impacts of Sea-Level Rise on Wave-Dominated Australian Estuaries with Differing Geomorphology. *J. Mar. Sci. Eng.* **2018**, *6*, 66. [CrossRef]

9.   Hsu, T.-W.; Shih, D.-S.; Li, C.-Y.; Lan, Y.-J.; Lin, Y.-C. A Study on Coastal Flooding and Risk Assessment under Climate Change in the Mid-Western Coast of Taiwan. *Water* **2017**, *9*, 390. [CrossRef]

10.   Masud, M.M.; Sackor, A.S.; Ferdous Alam, A.S.A.; Al-Amin, A.Q.; Ghani, A.B.A. Community responses to flood risk management—An empirical Investigation of the Marine Protected Areas (MPAs) in Malaysia. *Mar. Policy* **2018**, *97*, 119–126. [CrossRef]

11.   Vieira, B.F.V.; Pinho, J.L.S.; Barros, J.A.O.; Antunes do Carmo, J.S. Hydrodynamics and Morphodynamics Performance Assessment of Three Coastal Protection Structures. *J. Mar. Sci. Eng.* **2020**, *8*, 175. [CrossRef]

12.   Kron, W. Coasts: The high-risk areas of the world. *Nat. Hazards* **2013**, *66*, 1363–1382. [CrossRef]

13.   Storch, H.; Downes, N.K. A scenario-based approach to assess Ho Chi Minh City's urban development strategies against the impact of climate change. *Cities* **2011**, *28*, 517–526. [CrossRef]

14.   Fu, X.; Song, J. Assessing the Economic Costs of Sea Level Rise and Benefits of Coastal Protection: A Spatiotemporal Approach. *Sustainability* **2017**, *9*, 1495. [CrossRef]

15.   Jabareen, Y. Planning the resilient city: Concepts and strategies for coping with climate change and environmental risk. *Cities* **2013**, *31*, 220–229. [CrossRef]

16.   Scyphers, S.B.; Beck, M.W.; Furman, K.L.; Haner, J.; Josephs, L.I.; Lynskey, R.; Keeler, A.G.; Landry, C.E.; Powers, S.P.; Webb, B.M.; et al. A Waterfront View of Coastal Hazards: Contextualizing Relationships among Geographic Exposure, Shoreline Type, and Hazard Concerns among Coastal Residents. *Sustainability* **2019**, *11*, 6687. [CrossRef]

17.   Ito, T.; Setoguchi, T.; Miyauchi, T.; Ishii, A.; Watanabe, N. Sustainable Downtown Development for the Tsunami-Prepared Urban Revitalization of Regional Coastal Cities. *Sustainability* **2019**, *11*, 1020. [CrossRef]

18.   Wijaya, N.; Nitivattananon, V.; Shrestha, R.P.; Kim, S.M. Drivers and Benefits of Integrating Climate Adaptation Measures into Urban Development: Experience from Coastal Cities of Indonesia. *Sustainability* **2020**, *12*, 750. [CrossRef]

19.   Ge, Y.; Dou, W.; Liu, N. Planning Resilient and Sustainable Cities: Identifying and Targeting Social Vulnerability to Climate Change. *Sustainability* **2017**, *9*, 1394. [CrossRef]

20.   Xie, H.; He, Y.; Xie, X. Exploring the factors influencing ecological land change for China's Beijing-Tianjin-Hebei Region using big data. *J. Clean. Prod.* **2017**, *142*, 677–687. [CrossRef]

21.   Yao, G.; Xie, H. Rural spatial restructuring inecologically fragile mountainous areas of Southern China: A case study of Changgang Town, Jiangxi Province. *J. Rural Stud.* **2016**, *47*, 435–448. [CrossRef]

22.   Wang, P.; Yao, G.; Liu, G. Spatial evaluation of the ecological importance based on GIS for environmental management: A case study in Xingguo county of China. *Ecol. Indic.* **2015**, *51*, 3–12.

23.   Wong, K.; Zhang, Y.; Tsou, J.Y.; Li, Y. Assessing Impervious Surface Changes in Sustainable Coastal Land Use: A Case Study in Hong Kong. *Sustainability* **2017**, *9*, 1029. [CrossRef]

24.   Gittman, R.K.; Peterson, C.H.; Currin, C.A.; Joel Fodrie, F.; Piehler, M.F.; Bruno, J.F. Living shorelines can enhance the nursery role of threatened estuarine habitats. *Ecol. Appl.* **2016**, *26*, 249–263. [CrossRef]

25.   Scyphers, S.B.; Powers, S.P.; Heck, K.L.; Byron, D., Jr. Oyster reefs as natural breakwaters mitigate shoreline loss and facilitate fisheries. *PLoS ONE* **2011**, *6*, e22396. [CrossRef]

26.   Piazza, B.P.; Banks, P.D.; La Peyre, M.K. The Potential for Created Oyster Shell Reefs as a Sustainable Shoreline Protection Strategy in Louisiana. *Restor. Ecol.* **2005**, *13*, 499–506. [CrossRef]

27.   Smith, C.S.; Puckett, B.; Gittman, R.K.; Peterson, C.H. Living shorelines enhanced the resilience of saltmarshes to Hurricane Matthew (2016). *Ecol. Appl.* **2018**, *28*, 871–877. [CrossRef]

28.   Feagin, R.A.; Figlus, J.; Zinnert, J.C.; Sigren, J.; Martínez, M.L.; Silva, R.; Smith, W.K.; Cox, D.; Young, D.R.; Carter, G. Going with the flow or against the grain? The promise of vegetation for protecting beaches, dunes, and barrier islands from erosion. *Front. Ecol. Environ.* **2015**, *13*, 203–210. [CrossRef]

29.   Maun, M.A. *The Biology of Coastal Sand Dunes*; Oxford University Press: Oxford, UK, 2009; ISBN 978-0-19-857036-3.

30. Spalding, M.D.; Ruffo, S.; Lacambra, C.; Meliane, I.; Hale, L.Z.; Shepard, C.C.; Beck, M.W. The role of ecosystems in coastal protection: Adapting to climate change and coastal hazards. *Ocean Coast. Manag.* **2014**, *90*, 50–57. [CrossRef]

31. Reguero, B.G.; Beck, M.W.; Bresch, D.N.; Calil, J.; Meliane, I. Comparing the cost effectiveness of nature-based and coastal adaptation: A case study from the Gulf Coast of the United States. *PLoS ONE* **2018**, *13*, e0192132. [CrossRef]

32. Martínez, M.L.; Vázquez, G.; White, D.A.; Thivet, G.; Brengues, M. Effects of burial by sand and inundation by fresh- and seawater on seed germination of five tropical beach species. *Can. J. Bot.* **2002**, *80*, 416–424. [CrossRef]

33. Costanza, R.; Pérez-Maqueo, O.; Martinez, M.L.; Sutton, P.; Anderson, S.J.; Mulder, K. The Value of Coastal Wetlands for Hurricane Protection. *Ecol. Econ.* **2008**, *37*, 241–248. [CrossRef]

34. Salgado, K.; Martinez, M.L. Is ecosystem-based coastal defense a realistic alternative? Exploring the evidence. *J. Coast. Conserv.* **2017**, *21*, 837–848. [CrossRef]

35. Narayan, S.; Beck, M.W.; Wilson, P.; Thomas, C.J.; Guerrero, A.; Shepard, C.C.; Reguero, B.G.; Franco, G.; Ingram, J.C.; Trespalacios, D. The Value of Coastal Wetlands for Flood Damage Reduction in the Northeastern USA. *Sci. Rep.* **2017**, *7*, 9463. [CrossRef] [PubMed]

36. Sutton-Grier, A.E.; Wowk, K.; Bamford, H. Future of our coasts: The potential for natural and hybrid infrastructure to enhance the resilience of our coastal communities, economies and ecosystems. *Environ. Sci. Policy* **2015**, *51*, 137–148. [CrossRef]

37. Arkema, K.K.; Guannel, G.; Verutes, G.; Wood, S.A.; Guerry, A.; Ruckelshaus, M.; Kareiva, P.; Lacayo, M.; Silver, J.M. Coastal habitats shield people and property from sea-level rise and storms. *Nat. Clim. Chang.* **2013**, *3*, 1–6. [CrossRef]

38. Barbier, E.B.; Koch, E.W.; Silliman, B.R.; Hacker, S.D.; Wolanski, E.; Primavera, J.; Granek, E.F.; Polasky, S.; Aswani, S.; Cramer, L.A.; et al. Coastal ecosystem-based management with nonlinear ecological functions and values. *Science* **2008**, *319*, 321–323. [CrossRef]

39. Möller, I.; Kudella, M.; Rupprecht, F.; Spencer, T.; Paul, M.; van Wesenbeeck, B.K.; Wolters, G.; Jensen, K.; Bouma, T.J.; Miranda-Lange, M.; et al. Wave attenuation over coastal salt marshes under storm surge conditions. *Nat. Geosci.* **2014**, *7*, 727–731. [CrossRef]

40. Spalding, M.D.; McIvor, A.L.; Beck, M.W.; Koch, E.W.; Möller, I.; Reed, D.J.; Rubinoff, P.; Spencer, T.; Tolhurst, T.J.; Wamsley, T.V.; et al. Coastal Ecosystems: A Critical Element of Risk Reduction. *Conserv. Lett.* **2013**, *7*, 293–301. [CrossRef]

41. Narayan, S.; Beck, M.W.; Reguero, B.G.; Losada, I.J.; van Wesenbeeck, B.; Pontee, N.; Sanchirico, J.N.; Ingram, J.C.; Lange, G.-M.; Burks-Copes, K.A. The Effectiveness, Costs and Coastal Protection Benefits of Natural and Nature-Based Defenses. *PLoS ONE* **2016**, *11*, e0154735. [CrossRef]

42. Perez-Maqueo, O.; Martinez, M.L.; Sanchez-Barrandas, F.C.; Kolb, M. Assessing Nature-Based Coastal Protection against Disasters Derived from Extreme Hydrometeorological Events in Mexico. *Sustainability* **2018**, *10*, 1317. [CrossRef]

43. Chua, T.-E.; Bonga, D.; Bermas-Atrigenio, N. Dynamics of Integrated Coastal Management: PEMSEA's Experience. *Coast. Manag.* **2006**, *34*, 303–322. [CrossRef]

44. Duarte, C.M.; Losada, I.J.; Hendriks, I.E.; Mazarrasa, I.; Marbà, N. The role of coastal plant communities for climate change mitigation and adaptation. *Nat. Clim. Chang.* **2013**, *3*, 961–968. [CrossRef]

45. McLeod, K.L.; Lubchenco, J.; Palumbi, S.; Rosenberg, A.A. *Scientific Consensus Statement on Marine Ecosystem-Based Management*; Communication Partnership for Science and the Sea; Duke University: Durham, NC, USA, 2005.

46. Bridges, T.S.; Wagner, P.W.; Burks-Copes, K.A.; Bates, M.E.; Collier, Z.A.; Fischenich, J.C.; Gailani, J.Z.; Leuck, L.D.; Piercy, C.D.; Rosati, J.D.; et al. *Use of Natural and Nature-Based Features (NNBF) for Coastal Resilience*; The US Army Engineer Research and Development Center (ERDC): Vicksburg, MS, USA, 2015; pp. 1–479.

47. Palmer, M.A.; Liu, J.; Matthews, J.H.; Mumba, M.; D'Odorico, P. Manage water in a green way. *Science* **2015**, *349*, 584–585. [CrossRef] [PubMed]

48. Temmerman, S.; Kirwan, M.L. Building land with a rising sea. *Science* **2015**, *349*, 588–589. [CrossRef]

49. Van der Nat, A.; Vellinga, P.; Leemans, R.; van Slobbe, E. Ranking coastal flood protection designs from engineered to nature-based. *Ecol. Eng.* **2016**, *87*, 80–90. [CrossRef]

50. Scyphers, S.B.; Powers, S.P.; Heck, K.L. Ecological value of submerged breakwaters for habitat enhancement on a residential scale. *Environ. Manag.* **2014**, *55*, 383–391. [CrossRef]

51. Ridge, J.T.; Rodriguez, A.B.; Fodrie, F.J.; Lindquist, N.L.; Brodeur, M.C.; Coleman, S.E.; Grabowski, J.H.; Theuerkauf, E.J. Maximizing oyster-reef growth supports green infrastructure with accelerating sea-level rise. *Sci. Rep.* **2015**, *5*, 14785. [CrossRef]

52. French, P. *Coastal Defenses*, 1st ed.; Routledge: London, UK, 2001; pp. 51–301.

53. Lee, W.-D.; Yoo, Y.-J.; Jeong, Y.-M.; Hur, D.-S. Experimental and Numerical Analysis on Hydraulic Characteristics of Coastal Aquifers with Seawall. *Water* **2019**, *11*, 2343. [CrossRef]

54. Contestabile, P.; Crispino, G.; Russo, S.; Gisonni, C.; Cascetta, F.; Vicinanza, D. Crown Wall Modifications as Response to Wave Overtopping under a Future Sea Level Scenario: An Experimental Parametric Study for an Innovative Composite Seawall. *Appl. Sci.* **2020**, *10*, 2227. [CrossRef]

55. Argente, G.; Gómez-Martín, M.E.; Medina, J.R. Hydraulic Stability of the Armor Layer of Overtopped Breakwaters. *J. Mar. Sci. Eng.* **2018**, *6*, 143. [CrossRef]

56. Iuppa, C.; Contestabile, P.; Cavallaro, L.; Foti, E.; Vicinanza, D. Hydraulic Performance of an Innovative Breakwater for Overtopping Wave Energy Conversion. *Sustainability* **2016**, *8*, 1226. [CrossRef]

57. Gomes, A.; Pinho, J.L.S.; Valente, T.; Antunes do Carmo, J.S.; Hegde, V.A. Performance Assessment of a Semi-Circular Breakwater through CFD Modelling. *J. Mar. Sci. Eng.* **2020**, *8*, 226. [CrossRef]

58. Lee, B.W.; Seo, J.; Park, W.-S.; Won, D. A Hydraulic Experimental Study of a Movable Barrier on a Revetment to Block Wave Overtopping. *Appl. Sci.* **2020**, *10*, 89. [CrossRef]

59. Chybowski, L.; Grządziel, Z.; Gawdzińska, K. Simulation and Experimental Studies of a Multi-Tubular Floating Sea Wave Damper. *Energies* **2018**, *11*, 1012. [CrossRef]

60. Nishold, S.S.P.; Sundaravadivelu, R.; Saha, N. Physical model study on geo-tube with gabion boxes for the application of coastal protection. *Arab. J. Geosci.* **2019**, *12*, 164. [CrossRef]

61. Cherkasova, L. Application of gabions for strengthening marine coastal slopes. *J. Phys. Conf. Ser.* **2019**, *1425*, 012206. [CrossRef]

62. Chen, Y.; Tang, X.; Zhan, L. *Advances in Environmental Geotechnics*, 1st ed.; Springer: Berlin, Germany, 2009; pp. 805–920.

63. Pilarczyk, L. *Dikes and Revetments*, 1st ed.; A.A. Balkema: Rotterdam, The Netherlands, 1998.

64. Kerpen, N.B.; Schoonees, T.; Schlurmann, T. Wave Overtopping of Stepped Revetments. *Water* **2019**, *11*, 1035. [CrossRef]

65. Kerpen, N.B.; Schoonees, T.; Schlurmann, T. Wave Impact Pressures on Stepped Revetments. *J. Mar. Sci. Eng.* **2018**, *6*, 156. [CrossRef]

66. Wu, Y.; Dai, H.; Wu, J. Comparative Study on Influences of Bank Slope Ecological Revetments on Water Quality Purification Pretreating Low-Polluted Waters. *Water* **2017**, *9*, 636. [CrossRef]

67. Ware, D.; Buckwell, A.; Tomlinson, R.; Foxwell-Norton, K.; Lazarow, N. Using Historical Responses to Shoreline Change on Australia's Gold Coast to Estimate Costs of Coastal Adaptation to Sea Level Rise. *J. Mar. Sci. Eng.* **2020**, *8*, 380. [CrossRef]

68. Hamza, W.; Tomasicchio, G.R.; Ligorio, F.; Lusito, L.; Francone, A. A Nourishment Performance Index for Beach Erosion/Accretion at Saadiyat Island in Abu Dhabi. *J. Mar. Sci. Eng.* **2019**, *7*, 173. [CrossRef]

69. Cox, R.; Jahn, K.; Watkins, O.; Cox, P. Extraordinary boulder transport by storm waves (west of Ireland, winter 2013–2014), and criteria for analyzing coastal boulder deposit. *Earth-Sci. Rev.* **2018**, *177*, 623–636. [CrossRef]

70. Ávila, S.P.; Johnson, M.E.; Rebelo, A.C.; Baptista, L.; Melo, C.S. Comparison of Modern and Pleistocene (MIS 5e) Coastal Boulder Deposits from Santa Maria Island (Azores Archipelago, NE Atlantic Ocean). *J. Mar. Sci. Eng.* **2020**, *8*, 386. [CrossRef]

71. Othman, M. Value of mangroves in coastal protection. *Hydrobiologia* **1994**, *285*, 277–282. [CrossRef]

72. Verhagen, H.J. Financial Benefits of Mangroves for Surge Prone High-Value Areas. *Water* **2019**, *11*, 2374. [CrossRef]

73. Ma, C.; Ai, B.; Zhao, J.; Xu, X.; Huang, W. Change Detection of Mangrove Forests in Coastal Guangdong during the Past Three Decades Based on Remote Sensing Data. *Remote Sens.* **2019**, *11*, 921. [CrossRef]

74. Gilman, E.; Ellison, J.; Duke, N.; Field, C. Threats to mangroves from climate change and adaptation options. A review. *Aquat. Bot.* **2008**, *89*, 237–250. [CrossRef]

75. Angeletti, L.; Taviani, M. Offshore Neopycnodonte Oyster Reefs in the Mediterranean Sea. *Diversity* **2020**, *12*, 92. [CrossRef]

76. Zhao, M.; Zhang, H.; Zhong, Y.; Jiang, D.; Liu, G.; Yan, H.; Zhang, H.; Guo, P.; Li, C.; Yang, H.; et al. The Status of Coral Reefs and Its Importance for Coastal Protection: A Case Study of Northeastern Hainan Island, South China Sea. *Sustainability* **2019**, *11*, 4354. [CrossRef]

77. Christianen, M.; van Belzen, J.; Herman, P.; van Katwijk, M.; Lamers, L.; van Leent, P.; Bouma, T. Low-canopy seagrass beds still provide important coastal protection. *PLoS ONE* **2013**, *8*, e62413. [CrossRef] [PubMed]

78. Ondiviela, B.; Losada, I.; Lara, J.; Maza, M.; Galvan, C.; Bouma, T.; van Belzen, J. The role of seagrasses in coastal protection in a changing climate. *Coast. Eng.* **2014**, *87*, 158–168. [CrossRef]

79. Saponieri, A.; Valentini, N.; Di Risio, M.; Pasquali, D.; Damiani, L. Laboratory Investigation on the Evolution of a Sandy Beach Nourishment Protected by a Mixed Soft–Hard System. *Water* **2018**, *10*, 1171. [CrossRef]

80. Herbert, D.; Astrom, E.; Bersoza, A.C.; Batzer, A.; McGovern, P.; Angelini, C.; Wasman, S.; Dix, N.; Sheremet, A. Mitigating Erosional Effects Induced by Boat Wakes with Living Shorelines. *Sustainability* **2018**, *10*, 436. [CrossRef]

81. Muñoz-Perez, J.J.; Gallop, S.L.; Moreno, L.J. A Comparison of Beach Nourishment Methodology and Performance at Two Fringing Reef Beaches in Waikiki (Hawaii, USA) and Cadiz (SW Spain). *J. Mar. Sci. Eng.* **2020**, *8*, 266. [CrossRef]

82. Bitan, M.; Zviely, D. Sand Beach Nourishment: Experience from the Mediterranean Coast of Israel. *J. Mar. Sci. Eng.* **2020**, *8*, 273. [CrossRef]

83. Escudero, M.; Mendoza, E.; Silva, R. Micro Sand Engine Beach Stabilization Strategy at Puerto Morelos, Mexico. *J. Mar. Sci. Eng.* **2020**, *8*, 247. [CrossRef]

84. French, P. Managed retreat: A natural analogue from the Medway estuary, UK. *Ocean Coast. Manag.* **1999**, *42*, 49–62. [CrossRef]

85. Lawrence, J.; Bell, R.; Stroombergen, A. A Hybrid Process to Address Uncertainty and Changing Climate Risk in Coastal Areas Using Dynamic Adaptive Pathways Planning, Multi-Criteria Decision Analysis & Real Options Analysis: A New Zealand Application. *Sustainability* **2019**, *11*, 406.

86. Hanna, C.; White, I.; Glavovic, B. The Uncertainty Contagion: Revealing the Interrelated, Cascading Uncertainties of Managed Retreat. *Sustainability* **2020**, *12*, 736. [CrossRef]

87. Allen, J.R.L. Morphodynamics of holocene salt marshes: A review sketch from the atlantic and Southern North Sea coasts of Europe. *Quat. Sci. Rev.* **2000**, *19*, 1155–1231. [CrossRef]

88. Ahmadian, A. *Numerical Models for Submerged Breakwaters*, 1st ed.; Butterworth-Heinemann: Oxford, UK, 2016.

89. Ravindar, R.; Sriram, V.; Schimmels, S.; Stagonas, D. Characterization of breaking wave impact on vertical wall with recurve. *ISH J. Hydraul. Eng.* **2017**, *25*, 153–161. [CrossRef]

90. Castellino, M.; Lara, J.L.; Romano, A.; Losada, I.J.; De Girolamo, P. Wave loading for recurved parapet walls in non-breaking wave conditions: Analysis of the induced impulsive forces. *Coast. Eng. Proc.* **2018**, *1*, 34. [CrossRef]

91. Castellino, M.; Sammarco, P.; Romano, A.; Martinelli, L.; Ruol, P.; Franco, L.; De Girolamo, P. Large impulsive forces on recurved parapets under non-breaking waves. A numerical study. *Coast. Eng.* **2018**, *136*, 1–15. [CrossRef]

92. Cooper, J.; Mckenna, J. Working with natural processes: The challenge for coastal protection strategies. *Geogr. J.* **2008**, *174*, 315–331. [CrossRef]

93. Rubinato, M.; Nichols, A.; Peng, Y.; Zhang, J.; Lashford, C.; Cai, Y.; Lin, P.; Tait, S. Urban and river flooding: Comparison of flood risk management approaches in the UK and China and an assessment of future knowledge needs. *Water Sci. Eng.* **2019**, *12*, 274–283. [CrossRef]

94. Rubinato, M.; Luo, M.; Zheng, X.; Shao, S. Advances in modelling and prediction on the impact of human activities and extreme events on environments. *Water* **2020**, *12*, 1768. [CrossRef]

95. Pilkey, O.H.; Davis, T.W. An analysis of coastal recession models, North Carolina coast. In *Sea-Level Fluctuation and Coastal Evolution*; Nummendal, D., Pilkey, O.H., Howard, J.D., Eds.; SEPM (Society for Sedimentary Geology); Special Publications: Tulsa, OK, USA, 1987; Volume 41, pp. 59–68.

96. Martinez-Grana, A.M.; Boski, T.; Goy, J.L.; Zazo, C.; Dabrio, C.J. Coastal-flood risk management in central Algarve: Vulnerability and flood risk indices (South Portugal). *Ecol. Indic.* **2016**, *71*, 302–316. [CrossRef]

97. Nichols, A.; Rubinato, M.; Cho, Y.H.; Wu, J. Optimal use of titanium dioxide colourant to enable water surfaces to be measured by Kinect Sensors. *Sensors* **2020**, *20*, 3507. [CrossRef]

98. Rojas, S.; Rubinato, M.; Nichols, A.; Shucksmith, J. Cost effective measuring technique to simultaneously quantify 2D velocity fields and depth-averaged solute concentrations in shallow water flows. *J. Flow Meas. Instrum.* **2018**, *64*, 213–223. [CrossRef]

99. Martins, R.; Rubinato, M.; Kesserwani, G.; Leandro, J.; Djordjevic, S.; Shucksmith, J. On the characteristics of velocity fields on the vicinity of manhole inlet grates during flood events. *Water Resour. Res.* **2018**, *54*, 6408–6422. [CrossRef]

100. Rubinato, M.; Seungsoo, L.; Martins, R.; Shucksmith, J. Surface to sewer flow exchange through circular inlets during urban flood conditions. *J. Hydroinform.* **2018**, *20*, 564–576. [CrossRef]

101. Phantuwongraj, S.; Choowong, M.; Nanayama, F.; Hisada, K.I.; Charusiri, P.; Chutakositkanon, V.; Pailoplee, S.; Chabangbon, A. Coastal geomorphic conditions and styles of storm surge washover deposit from Southern Thailand. *Geomorphology* **2013**, *192*, 43–58. [CrossRef]

102. Claudino-Sales, V.; Wang, P.; Horwitz, M.H. Factors controlling the survival of coastal dunes during multiple hurricane impacts in 2004 and 2005: Santa Rosa barrier island, Florida. *Geomorphology* **2008**, *95*, 295–315. [CrossRef]

103. Wang, P.; Horwitz, M.H. Erosional and depositional characteristics of regional overwash deposits caused by multiple hurricanes. *Sedimentology* **2007**, *54*, 545–564. [CrossRef]

104. Naylor, L.A.; Spencer, T.; Lane, S.N.; Darby, S.E.; Magiligan, F.J.; Macklin, M.G.; Moller, I. Stormy geomorphology: Geomorphic contributions in an age of climate extremes. *Earth Surf. Process. Landf.* **2017**, *42*, 166–190. [CrossRef]

105. Bernatchez, P.; Fraser, C.; Lefaivre, D.; Dugas, S. Integrating anthropogenic factors, geomorphological indicators and local knowledge in the analysis of coastal flooding and erosion hazards. *Ocean Coast. Manag.* **2011**, *54*, 621–632. [CrossRef]

106. Wu, S.; Rubinato, M.; Gui, Q. SPH Simulation of interior and exterior flow field characteristics of porous media. *Water* **2020**, *12*, 918. [CrossRef]

107. Zhang, Y.; Rubinato, M.; Kazemi, E.; Pu, J.H.; Huang, Y.; Lin, P. Numerical and experimental analysis of shallow turbulent flow over complex roughness beds. *Int. J. Comput. Fluid Dyn.* **2019**, *33*, 202–221. [CrossRef]

108. Shu, A.; Duan, G.; Rubinato, M.; Tian, L.; Wang, M.; Wang, S. An Experimental Study on Mechanisms for Sediment Transformation Due to Riverbank Collapse. *Water* **2019**, *11*, 529. [CrossRef]

109. Xie, L.; Liu, H.; Peng, M. The effect of wave-current interactions on the storm surge and inundation in Charleston Harbor during Hurricane Hugo, 1989. *Ocean Model.* **2008**, *20*, 252–269. [CrossRef]

110. Donelan, W.A.; Dobson, F.W.; Smith, S.D. On the dependence of sea surface roughness on wave development. *J. Phys. Oceanogragr.* **1993**, *23*, 2143–2149. [CrossRef]

111. Mellor, G.L. The three-dimensional current and surface wave equations. *J. Phys. Oceanogragr.* **2003**, *33*, 1978–1989. [CrossRef]

112. Komen, G.J.; Cavaleri, L.; Donelan, M.; Hasselmann, K.; Hasselmann, S.; Jansenn, P.A.E.M. *Dynamics and Modelling of Ocean Waves*; Cambridge University Press: New York, NY, USA, 1994; p. 532.

113. Lin, R.Q.; Huang, N.E. The Goddart coastal wave model, 1, Numerical method. *J. Phys. Oceanogragr.* **1996**, *26*, 833–847. [CrossRef]

114. Jokiel, P.L. Impact of storm waves and storm floods on Hawaiian reefs. In Proceedings of the 10th International Coral Reef Symposium, Okinawa, Japan, 28 June–2 July 2006; pp. 282–284.

115. Carter, R.W.G.; Lowry, P.; Stone, G.W. Sub-tidal ebb-shoal control of shoreline erosion via wave refraction, Magilligan Foreland, Northern Ireland. *Mar. Geol.* **1982**, *48*, M17–M25. [CrossRef]

116. Lyddon, C.E.; Brown, J.M.; Leonardi, N.; Saulter, A.; Plater, A.J. Quantification of the uncertainty in coastal storm hazard predictions due to wave-current interaction and wind forcing. *Geophys. Res. Lett.* **2019**, *46*. [CrossRef]

Article

Descriptive Analysis of the Performance of a Vegetated Swale through Long-Term Hydrological Monitoring: A Case Study from Coventry, UK

Luis A. Sañudo-Fontaneda [1,2,*], Jorge Roces-García [1], Stephen J. Coupe [2], Esther Barrios-Crespo [1], Carlos Rey-Mahía [1], Felipe P. Álvarez-Rabanal [1] and Craig Lashford [2]

[1]   INDUROT Research Institute, GICONSIME Research Group, Department of Construction and Manufacturing Engineering, University of Oviedo, Campus of Mieres, Gonzalo Gutierrez Quiros s/n, 33600 Mieres, Spain; rocesjorge@uniovi.es (J.R.-G.); uo245409@uniovi.es (E.B.-C.); UO236881@uniovi.es (C.R.-M.); alvarezfelipe@uniovi.es (F.P.Á.-R.)

[2]   Centre for Agroecology, Water and Resilience, Coventry University, Ryton Gardens, Coventry CV8 3LG, UK; Stephen.coupe@coventry.ac.uk (S.J.C.); ab0874@coventry.ac.uk (C.L.)

*   Correspondence: sanudoluis@uniovi.es; Tel.: +3-49-8545-8196

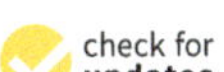

Received: 15 August 2020; Accepted: 30 September 2020; Published: 6 October 2020

Abstract: Vegetated swales are a popular sustainable drainage system (SuDS) used in a wide range of environments from urban areas and transport infrastructure, to rural environments, sub-urban and natural catchments. Despite the fact that vegetated swales, also known as grassed swales, have received scientific attention over recent years, especially from a hydrological perspective, there is a need for further research in the field, with long-term monitoring. In addition, vegetated swales introduce further difficulties, such as the biological growth occurring in their surface layer, as well as the biological evolution taking place in them. New developments, such as the implementation of thermal devices within the cross-section of green SuDS for energy saving purposes, require a better understanding of the long-term performance of the surface temperature of swales. This research aims to contribute to a better understanding of these knowledge gaps through a descriptive analysis of a vegetated swale in Ryton, Coventry, UK, under a Cfb Köppen climatic classification and a mixed rural and peri-urban scenario. Precipitation and temperature patterns associated with seasonality effects were identified. Furthermore, a level of biological evolution was described due to the lack of periodical and planned maintenance activities, reporting the presence of both plant species and pollinators. Only one event of flooding was identified during the three hydrological years monitored in this research study, showing a robust performance.

Keywords: biological evolution; ecosystem services; low impact development (LID); stormwater best management practices (BMP); stormwater control measures (SCMs); sustainable drainage systems (SuDS); water sensitive urban design (WSUD)

1. Introduction

Sustainable drainage systems (SuDS) are nature-based solutions (NBS) utilised to manage water, both in urban and rural environments, as well as in transport infrastructures. They are often referred to as stormwater best management practices (BMP), water sensitive urban design (WSUD), stormwater control measures (SCM) and low impact developments (LID), amongst other terminology [1].

Swales are SuDS that are mainly utilised in transportation infrastructure and in urban and sub-urban environments to capture pollutants and attenuate runoff volumes [2–5]. Furthermore, they are used in rural environments and farms to manage stormwater [6]. These techniques also provide landscape features, as well as an improvement in biodiversity and amenity [7]. In addition, swales have been utilised in permaculture practices showing a robust long-term performance, as highlighted by Abrahams et al. [8]. These authors, along with Winston et al. [9], related the ecosystem services provided by swales, to those delivered by wetlands, especially when vegetation growth is allowed under a low maintenance condition. In other words, allowing nature to take ownership of the system up to some degree.

Vegetated swales have been treated in scientific analyses as conventional or standard swales, as indicated by Fardel et al. [10]. Therefore, they were included in the same category as swales, grassy swales, vegetated roadside swales, planted swales and grassy media, amongst others. Other categories refer to dry, wet and bio-swales. However, dry swales are often described as swales able to completely drain stormwater runoff between two consecutive storm events by authors such as Hunt et al. [11], which also includes some vegetated swales in this category.

Fardel et al. [10] gathered the main parameters influencing swale performance in the literature up to 2019. The authors distinguished between those variables affecting the drainage area, such as the discharge area, the discharge ratio and the main concentration at the swale inlet; those variables associated with the swale itself, such as the swale length, slope, type of soil, vegetation and operational life. This research also revealed the limitations of previous work, emphasizing the need for continuous and long-term monitoring alongside later work such as Purvis et al. [12]. Most of the investigations carried out in the literature show a limited number of storm events which introduces a certain level of uncertainty, as they miss the major hydrological effects influenced by the climate on the location, the rainfall and temperature patterns and the seasonality. To minimise this problem, McCarthy et al. [13] proposed a minimum range of 15–20 storm events in order to capture robust data from a water quality stand point. Therefore, the hydrology should also reach this threshold in order to be reliable and representative for comparison with other international studies. Recent research, such as Purvis et al. [12], monitored 39 storm events over 12 months in a bio-swale in North Carolina, USA, which also followed the described conditions.

Temperature relationships within swales were highlighted as an important factor, especially when considering potential energy applications like the implementation of a ground source heat pump (GSHP) as outlined by Charlesworth et al. and Rey-Mahía et al. [14,15]. Both reports indicated that more research is needed to understand the long-term patterns and their role in the hydrological impact on pervious pavements and vegetated swales.

The most up-to-date reviews and scientific research on vegetated swales, such as Gavrić et al. [16], pointed out the need for improved modelling in grass–soil media, being underpinned by a better knowledge of physical processes taking place in this SuDS technique. Furthermore, complete facility descriptions ideally are required, to fully describe the functions and ecosystem services provided by vegetated swales.

Design guidance for new developments should specify the implementation of SuDS treatment trains for stormwater management, based on recent studies carried out by Williams et al. [17] regarding user perception of SuDS benefits. Additionally, treatment trains can connect to further social and ecological elements of the urban landscape, due to their comprehensive and holistic design features, as pointed out by Lähde et al. [18]. Treatment trains containing green roofs and grassed swales promote hydrological processes of detention and conveyance, including infiltration within the swale, when designed for that purpose [19].

Once the knowledge gaps were identified in the literature, the research presented in this paper aimed to provide further information about the long-term hydrological processes, occurring in a vegetated swale performing under real weather conditions in the field, showing the seasonality effect as well as the evolution of the water temperature of the system. This research also highlights the hydrological impact of a vegetated swale within a treatment train, when associated with an extensive green roofs.

The specific objectives of this research are cited as follows:

- To accurately describe the hydrological patterns through water level measurement at the discharge point, identifying levels of potential hydraulic failure as well as the storm events that may cause any failure, over three hydrological years.
- To identify surface temperature trends in the vegetated swale over several years of operation.
- To discuss the results depending on the maintenance activities carried out over the research period and the biological evolution of the vegetated swale.

Long-term hydrological and temperature monitoring alongside operation and maintenance monitoring allows detailed investigation of the performance of vegetated swales in the field, to develop a better understanding of this SuDS technique.

2. Materials and Methods

The experimental methods utilised in this research are presented in the following sub-sections, including monitoring equipment, hypotheses, materials and methods.

2.1. Location and Climate Context

The experimental site is located within the premises of the Centre for Agroecology, Water and Resilience (CAWR), Coventry University (52°37′ N, 1°41′ W), in Ryton Organic Gardens, Ryton-on-Dunsmore, Warwickshire, UK. The site, constructed in 2005, is 1 km away from the A45 highway and 1.5 km from the village of Wolston (Figure 1). The land use of the study area could be defined as rural mixed with small peri-urban areas associated with roads, highways, car parks and other civil engineering related infrastructure, as well as small villages. The organic gardens also contain other SuDS devices across the complex such as filter strips, a reed bed, rain gardens and large green areas promoting infiltration and bioretention hydrological processes.

Figure 1. Study area in Ryton-on-Dunsmore, Coventry, and location of the UK Met Office weather stations.

Coventry has a Cfb climate with warm temperature, fully humid and warm summers, based on the Köppen–Geiger climatic classification used to categorise the climate conditions across the World. Historical weather data for Coventry and its surrounding area have been obtained for the historical series between 1981 and 2010, showing 700.1 mm annual rainfall with 124 days/year with precipitation over 1 mm for Coundon, Coventry, comparing with the UK average values for annual rainfall (1154 mm) and days of rainfall over 1 mm per annum (156) [20]. Data from Church Lawford for the same period exhibited 674.8 mm annual rainfall and 121 days/year of rainfall over 1 mm. In addition, Coleshill presented 712.4 mm annual rainfall and 129 days/year of rainfall over 1 mm. Finally, Wellesbourne showed 614.8 mm annual rainfall and 114 days/year of rainfall over 1 mm (see Table 1).

Table 1. Historical weather data for Coventry and its surrounding area, ranging between 1981 and 2010 [20].

Station Name	Latitude (N)	Longitude (W)	Altitude above Mean Sea Level (m)	Distance from City Centre (km)	Station Type
Coventry, Coundon	52.42	1.53	119	3.2	Observing site
Church Lawford	52.36	1.33	107	12.9	Observing site
Coleshill	52.48	1.69	96	14.5	Observing site
Wellesbourne	52.21	1.60	47	22.5	Observing site

It is important to put into context for this study, that climate change effects related to temperatures registered in the UK, especially affected the monitoring period of this study. The ten warmest years in the UK since 1884 have taken place between 2002 and 2019 as reported by the Met Office and also emphasised by the British Broadcasting Corporation (BBC) [21]. This time range covers the full period of monitoring for this field study, highlighting that the 7th warmest year on record was 2018, with July 2019 breaking the record for the hottest day ever recorded in the UK (38.7 °C in Cambridge).

2.2. Case Study: Swale and SuDS Treatment Train

The vegetated swale monitored in this research is part of a wider drainage system based in a SuDS treatment train, constructed in 2005, including an extensive green roof which drained into the swale through 3 downspouts (Figure 2). Then, this study could be considered as one of the reports presenting a larger service life for a swale, in an international study so far, considering the investigation of swales reported by Fardel et al. [10] in their review. The area drained by the swale consists of several differentiated sub-catchments depicted in Figure 2, with general characteristics presented in Table 2 as follows.

Table 2. Estimated areas of the sub-catchments of the drainage area and runoff coefficients from the contributing areas.

Sub-Catchment Type	Area (m^2)	Estimated Runoff Coefficient
Paved area	225.0	>0.90 [22]
Extensive green roof	150.0	0.40–0.70 [1] [23–25]
Vegetated swale	157.5	——

[1] Values associated to the sub-surface layers which limited the whole runoff coefficient for the roofs.

The vegetated swale is approximately 45.0 m length and 1.1 m wide at the bottom, with 1.0–1.5% slope, a trapezoidal cross-section, consisting of layers of natural soil and vegetation. The ponding zone is 0.6 m deep and 2:1 side slope (H:V) for a total of around 3.5 m width. Two 40 mm diameter pipes are placed at the discharge point which is further connected downstream to the River Avon.

The contributing area is 375.0 m^2 (Table 2), divided into an impervious paved area (225.0 m^2) and an extensive green roof (150.0 m^2). The ratio between the contributing area and the vegetated swale is 2.4:1 with 42.3% of catchment being impervious (Table 2). These values are within the range of the usual swale area presented by Fardel et al. [10] in their study covering 59 swales internationally.

The average infiltration rate for a grassed swale is considered to be around 13 mm/h based on Ariza et al. [19] whilst the minimum threshold recommended is 2.7 mm/h by the Maryland Department of Environment [26]. The infiltration for the swale was characterised using a system to determine the time to drain the temporary ponding water under a no rainfall scenario, clogging of the discharge point and different water levels (Table 3). It is important to note that the values registered in Table 3 do not differentiate between evapotranspiration and infiltration, therefore accounting for both factors.

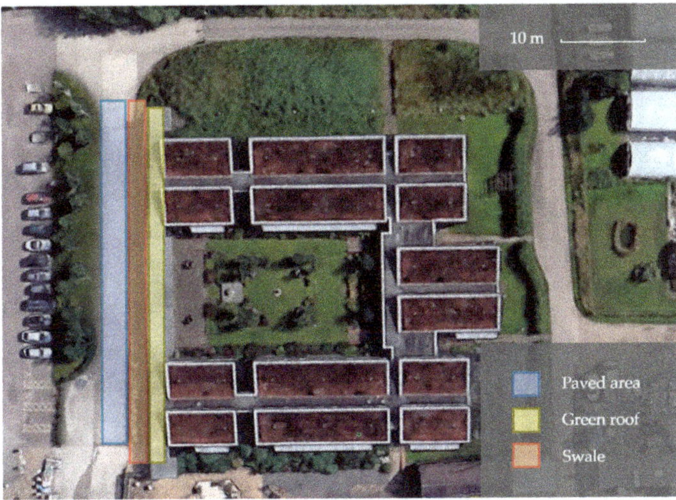

Figure 2. Drainage area and downspouts draining the runoff from the adjacent extensive green roof.

Table 3. Determination of the average time to empty the temporary ponding in the vegetated swale.

Date	Initial Water Level [cm]	Final Water Level [cm]	Time [h]	Empty Rate [mm/h]
4-2-19 17:00	8.0	3.0	25	2.00
9-2-19 3:00	10.0	5.0	22	2.27
10-2-19 17:00	14.0	0.0	182	0.77
1-3-19 6:00	7.0	1.0	49	1.24
4-3-19 0:00	6.0	4.0	14	1.43
6-3-19 21:00	14.0	11.0	10	3.00
10-3-19 21:00	12.0	9.0	17	1.76
12-3-19 17:00	22.0	13.0	25	3.60
2-4-19 19:00	5.0	1.0	40	1.01
2-5-19 22:00	1.0	0.0	5	2.00
8-5-19 18:00	10.0	7.0	6	5.00
7-6-19 18:00	5.0	2.0	7	4.29
8-6-19 13:00	11.0	0.0	24	4.58
10-6-19 0:00	5.0	2.0	8	3.75
12-6-19 10:00	9.0	7.0	6	3.37
13-6-19 12:00	13.0	11.0	5	4.00
15-6-19 18:00	11.0	4.0	52	1.35
19-6-19 1:00	7.0	0.0	63	1.11
25-6-19 14:00	12.0	0.0	94	1.28

A total of 19 events were analysed in the latest stage of the monitoring period presented in this research, resulting in an average ratio of 2.52 mm/h to empty the temporary ponding water in the vegetated swale. This data are close to that recommended by stormwater design manuals such as the Maryland Department of Environment [26].

2.3. Hydrological and Temperature Monitoring

Water level and surface temperature were registered at the discharge point of the vegetated swale, using an Orpheus Mini pressure probe for ground water and open water level measurements with an integrated temperature sensor (OTT Hydromet). The resolution provided is 0.1 °C for temperature and 0.05% FS accuracy for pressure (percentage of full scale), including a ±0.1%/year FS long-term stability. Data were collected at a 1 h interval. The monitoring period for the study was defined between the 21 June 2016 to the 30 September 2019, covering 3.25 years and 3 hydrological years (period defined between 1 October of one year and 30 September of the next by the United States Geological Survey).

Water level monitoring allowed the authors to determine the performance of the swale from a hydrological perspective, by identifying the temporary ponding produced by different storm events over the 3 hydrological years. This temporary ponding enabled the detection of those rainfalls that produce sufficient runoff to exceed the hydraulic capacity of the system, and overflow. The water levels used to pinpoint the 2 levels of hydraulic failure were defined as the diameter of the pipes placed at the discharge point (0.40 m), whilst flooding issues were depicted as those values recorded over the maximum ponding depth (0.60 m) (Figure 3). Then, those water level records registered above these threshold values, would mean that the system failed to cope with the storm event under one of the two levels of security considered.

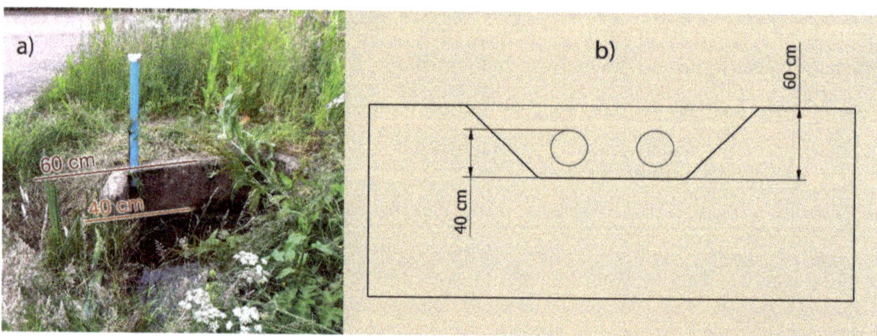

Figure 3. Water levels of hydraulic failure indicated at the discharge point (**a**) and its cross section (**b**).

Storm events were isolated by using 6 h slots where no rainfall was recorded over 2.5 mm values [27]. The reason for choosing 2.5 mm as a minimum threshold lies in the minimum values required to wet the soil during a storm event; the first 2.5 mm of rainfall are not accounted for runoff volume. Interception considerations for swales in the UK SuDS Manual indicate 5 mm as the minimum threshold to isolate storm events liable to create runoff or to contribute to water ponding in the system [7]. Thus, the 2.5 mm value was selected for being more restrictive for the hydrological analyses.

This research presents a descriptive analysis over the 3 year period, plus a more detailed analysis during the 6 months when, on-site weather data were available. The isolation of storm events was possible during that time. Precipitation and temperature data provided for the rest of the monitoring period were given at daily intervals.

The seasonality effect is identified and described through a long-term hydrological and temperature monitoring over 3 hydrological years.

2.4. Climate Data Collection

The following weather stations shown in Table 4 were selected to collect weather data for this study for the period of time starting at the 21 June 2016 and ending at the 30 September 2019 (see location in Figure 1). The on-site weather station was used over 6 months of the monitoring period beginning to register data by the 1 February and ending on the 31 August 2019. Temperature data were collected from Church Lawford and Coundon, Coventry, whilst precipitation was obtained from Church Lawford, Finham and Draycote weather stations (Table 4).

Table 4. Weather stations utilised in the study for the monitoring period of this research: 21 June 2019–30 September 2019. Source: UK Met Office.

Station Name	Latitude (N)	Longitude (W)	Altitude above Mean Sea Level (m)	Distance from the Site (km)	Weighting Coefficient (Rain/Temp)
Church Lawford	52.36	1.33	107	6.68	0.29/0.55
Coundon, Coventry	52.42	1.53	119	8.40	—/0.45
Finham	52.36	1.50	65	5.30	0.47/—
Draycote	52.32	1.32	90	7.45	0.24/—

A HOBO U30 NRC (Onset) weather station was used to collect weather data, such as wind speed, wind direction, solar radiation, relative humidity, barometric pressure, gust speed, dew point, air temperature and precipitation on-site. Precipitation and temperature were measured at a 1 min interval using a 0.2 mm resolution tipping-bucket rain gauge and a smart 12 bit temperature sensor, respectively.

Therefore, it is possible to distinguish 2 stages defined by the availability of the on-site weather station data. Weather data from the first 31 months were adjusted through the analysis carried out during the 6 months when data from all sources were available. The method utilised to calibrate the first 31 months followed the inverse distance weighted (IDW) interpolation method (Equation (1), which is described as a simplified way to obtain interpolate data when no weather station is available on-site [28]. More details are presented in Figure 1 regarding the location and Table 4 for the specific characteristics of the UK Met Office weather stations utilised in this method.

$$P_X = \sum_{i=1}^{n} \frac{P_i}{D_i^W} \Big/ \sum_{i=1}^{n} \frac{1}{D_i^W}, \quad W = 2, \tag{1}$$

where:

- P: Precipitation data.
- D: Distance.
- W: Power factor.

Weighting coefficients (Table 4) were calculated for each of the UK Met Office's weather station outputs, based on the accuracy of their values in relation to those available from the on-site weather station during the 6 months period with on-site data available.

This methodology provided the following calibrated weather data for the whole period of monitoring where the first 31 months followed the method of calculation showed above and the last 6 months used the data from the weather station located on-site (Figure 4).

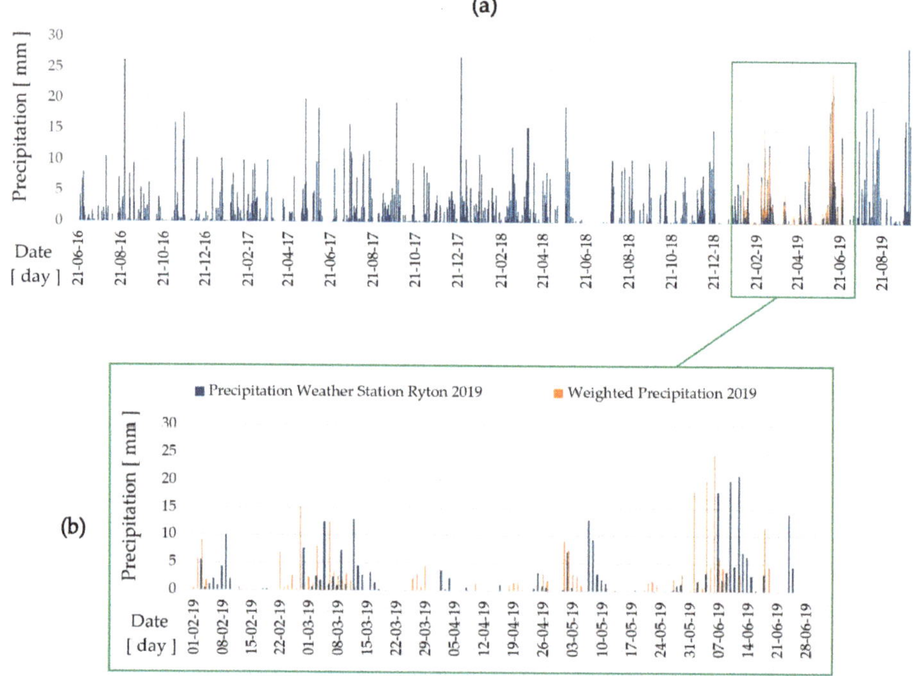

Figure 4. Weather data at the experimental site for (**a**) the entire monitoring period and (**b**) the 6 months of calibration.

2.5. Operation and Maintenance over the Research Period

Between 2016 and 2019, the swale received only superficial reactive maintenance, consisting primarily of an autumnal cut of large vegetation such as the removal of weeds >300 mm in height and coppicing of saplings that had established in the swale. This reduced the visible growth in the winter periods, with an annual regrowth occurring the following spring. A change of ownership of the site occurred in early 2020, resulting in a clear cut and complete removal of vegetation on 3 February 2020 (see Figure 5).

The maintenance schedule allowed the growth of vegetation (see Figure 6) which involved a certain degree of biological evolution, potentially transforming the ecosystem service provided by the system. The removal of large vegetation could prevent out-competition of beneficial plants, such as those that support pollinators, but could limit the hydraulic attenuation of the swale by reducing evapotranspiration and vegetative obstacles to flow.

Prior to the reactive maintenance, there was evidence that the swale had some biodiversity and ecosystem services value, with extensive bramble growth (*Rubus fruticosus*), goat willow (*Salix caprea*), bulrush (*Typha latifolia*), common nettle (*Urtica dioica*) and daffodil (*Narcissus*), as shown in Figure 5. Several species of grass were present, and the majority of the plants encountered were self-seeded, coming into the swale from vegetation adjacent to the swale. Authors noted several specimens of pollinating insect using the swale, including honeybees, bumble bees, hoverflies, butterflies and moths (Figure 6).

Figure 5. Detail of the swale after complete removal of vegetation (February 2020): (**a**) South-west sight of the swale; (**b**) North-east view of the swale from the discharge point and the monitoring device.

Figure 6. Detail of the swale before maintenance works. Picture taken looking from North-east to South-west from the mid-point of the swale's length in early Spring 2019.

3. Results

The results in this paper are divided into three main sub-sections according to the specific objectives of the research: hydrological performance, temperature behaviour and operation and maintenance.

3.1. Hydrological Performance

Three hydrological years were monitored for this study, showing varying storm events from low values to storm events with peaks over 25 mm. Figure 7 shows the direct impact of the storm events on the hydrology of the vegetated swale, by the increment in the water level at the discharge point.

It was possible to identify a repeated pattern in 2018 and 2019 for the water level in winter, especially between December and March where water levels ranged between 5 and 15 cm, with some peaks over 20 cm. However, the winter of 2019 shows an irregular performance with some valleys showing no detectable water level.

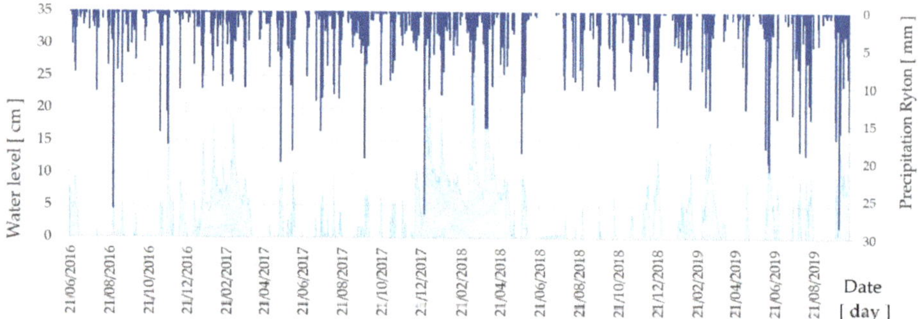

Figure 7. Water level and rainfall registered during this research.

Summers indicated lower water levels from 0 to 10 cm (Figure 7) and many consecutive days with no temporary water standing in the swale. This long-term descriptive analysis allows the identification of those periods where the vegetated swale varies from dry swale to wet swale, including the identification of maintenance operations marking changes between grassed swale functions and the provision of wet land services with relevant growing of vegetation which will be described in the operation and maintenance sub-section (3.3).

Individual storm events were identified and isolated following the methodology described in previously in the paper and presented in Table 5, as follows over a 6 month period, being similar to the monitoring periods used by other researchers such as Lisenbee et al. [29].

Table 5. Storm events captured during the 6 month period in 2019 and water level variations.

Start Date	End Date	Rain Accumulated [mm]	Initial Water Level [mm]	Final Water Level [mm]	Rain Rate [mm/h]	H [cm]
4-2-19 4:00	4-2-19 17:00	2.9	0.0	8.0	0.22	8.0
8-2-19 14:00	9-2-19 7:00	2.1	4.0	9.0	0.12	5.0
10-2-19 4:00	10-2-19 17:00	7.5	6.0	12.0	0.58	6.0
28-2-19 19:00	1-3-19 6:00	5.1	0.0	7.0	0.46	7.0
3-3-19 14:00	3-3-19 20:00	0.1	4.0	6.0	0.02	2.0
5-3-19 23:00	7-3-19 2:00	13.7	2.0	12.0	4.57	10.0
10-3-19 3:00	10-3-19 15:00	4.3	9.0	12.0	0.36	3.0
12-3-19 10:00	12-3-19 20:00	10.1	10.0	20.0	1.01	10.0
2-4-19 10:00	2-4-19 19:00	1.1	1.0	5.0	0.12	4.0
2-5-19 13:00	3-5-19 0:00	4.5	0.0	1.0	0.41	1.0
8-5-19 5:00	8-5-19 22:00	10.3	0.0	8.0	0.61	8.0
7-6-19 9:00	7-6-19 22:00	8.1	0.0	3.0	0.62	3.0
8-6-19 6:00	8-6-19 18:00	6.7	1.0	7.0	0.56	6.0
9-6-19 18:00	10-6-19 4:00	0.9	0.0	4.0	0.09	4.0
10-6-19 15:00	12-6-19 11:00	24.5	4.0	8.0	1.23	4.0
12-6-19 18:00	13-6-19 14:00	23.7	7.0	12.0	1.19	5.0
15-6-19 15:00	15-6-19 21:00	0.3	8.0	10.0	0.05	2.0
18-6-19 17:00	19-6-19 3:00	0.5	3.0	6.0	0.05	3.0
25-6-19 4:00	25-6-19 20:00	18.9	0.0	10.0	1.18	10.0

Maximum rainfall events were recorded around 25 mm, while the usual storm events were registered between 5 and 10 mm. The duration of the storm events lasted a few hours up to several days recording low intensity ratios (Table 5).

Table 6 presents those water levels registered in the vegetated swale that registered values over the threshold identified in the methodology, reaching one of the two levels or types of failure. Only one flooding issue was recorded during the 3 years of monitoring (November 2016). One hydraulic failure was identified in 2016. This pattern is repeated in 2017, with a further two events occurring in 2018 and none in 2019. This long-term monitoring allows the identification of potential emergency issues derived from high intensity storm events.

Table 6. Data for the water levels reaching one of the two levels of failure: (**a**) hydraulic failure (0.40 –0.60 m); (**b**) flooding issue (>0.60 m).

Date	Water Level (mm)	Type of Failure
21/11/2016	38	Hydraulic Failure
30/11/2016	68	Flooding Issue
27/12/2017	37	Hydraulic Failure
31/03/2018	37	Hydraulic Failure
02/04/2018	41	Hydraulic Failure

3.2. Temperature Behaviour

Seasonality effects were captured in the field monitoring, being especially clear in the descriptive analysis of the temperatures, recorded in the surface of the vegetated swale. Air temperature also followed the same trend, with peaks in summer and valleys in winter (Figure 8).

This temperature behaviour presented in Figure 8, reinforces the idea of the design of greener SuDS, such as devices housing GSHP technologies, reported by Charlesworth et al. [14]. Rey-Mahía et al. [15], in their laboratory simulation of the combination of swales and GSHP technologies, presented a range of temperature performance that can be compared to this field study in order to find out whether this laboratory study could be applied to the swale in Ryton, Coventry, UK. Figure 8 shows that even in the worst scenario, represented by air temperature falling below 0 °C several times over the 3 year period of the study (nearly −5 °C in early 2018 as the lowest temperature registered during the monitoring period), the surface temperature for the swale maintained consistent values above 0 °C, temperatures which would be expected to be even higher in the ground below the swale surface. Consequently, the GSHP system could work continuously. Following on from this potential application, the swale surface registered mean temperature values around 5 °C in winter and 17 °C in summer, providing future studies with valuable information to feed their simulations and modelling investigations.

Figure 8 shows that the difference in temperature between the surface of the swale, and the air is pronounced, with much greater extremes in the air temperature. This difference reaches 4–6 °C with peaks up to 8 °C in the summer period. On the other hand, the variation is positive towards the surface in winters, reaching 1–2 °C of difference, being between 4 and 6 °C in the most favourable cases. This is another point of interest for researchers regarding the consistency of the temperature variation between the air and the swale, considering the climatic conditions for this study. In addition, this data could be of particular relevance to local authorities looking to reduce the urban heat island (UHI) effect in urban environments through the implementation of SuDS techniques. In the case of this swale, the temperature behaviour provided by the system was consistent and robust. No high extreme values were registered for the temperature of the swale surface despite the fact that the UK had some of its warmer summers during this monitoring period, as described in Section 2.1.

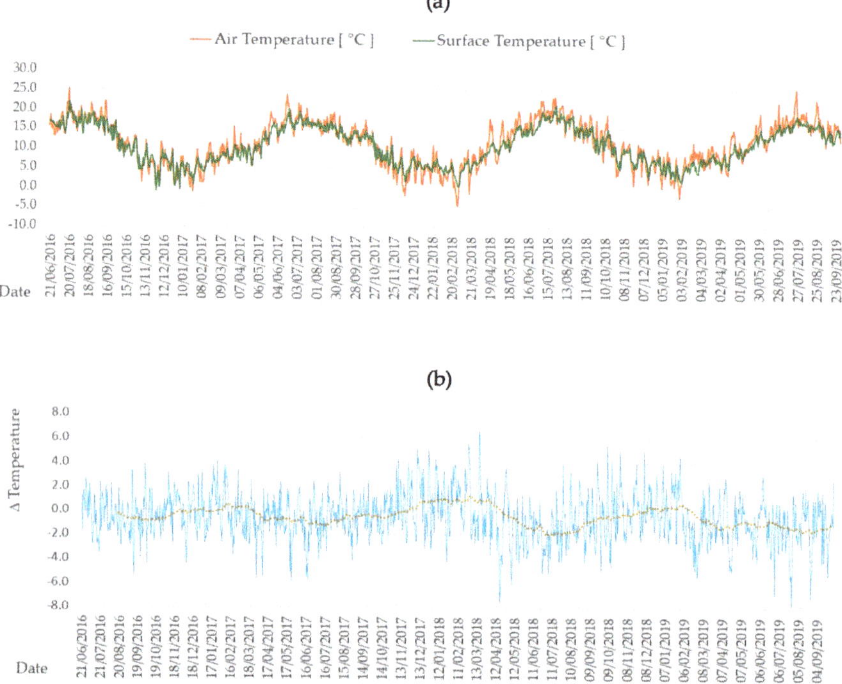

Figure 8. (**a**) Surface water temperature (°C) and air temperature (°C) registered during the monitoring period; (**b**) temperature variation between the surface and the air.

3.3. Operation and Maintenance

Operation and maintenance are key factors of the hydrological performance and pollutant removal efficiency of vegetated swales, as also reported by Horstmeyer et al. [30] and Johnson and Hunt [31] regarding other related popular stormwater practices in sub-urban areas such as bioretention.

Following on from Section 2 regarding maintenance, operation and maintenance activities—described in stormwater manuals and guides internationally such as Woods Ballard et al. [7], the Maryland Department of Environment [26], the North Carolina Department of Environmental Quality [32]—are compared with those carried out in the current research, putting this report into a wider context. The vegetative swale lacked any kind of short or medium term maintenance. Consequently, maintenance such as the removal of litter and debris, inspection of potential blockage in inlets and outlets, silt accumulation, vegetation coverage, removal of nuisance plants and grass cutting, amongst others, which are required on a monthly or annual basis, were not conducted.

No sediments were removed from the swale, resulting in a silt build-up of 2 cm at the discharge point over the entire period of monitoring. This silt accumulation was confirmed by the pressure sensor measuring the water level during dry periods with no water in the swale. The device was installed 4 cm over the surface of the swale, ending only at 2 cm over the surface, after 3 years of operation.

In conclusion, all maintenance was related to remedial activities after a problem was identified and was usually carried out once a year.

This research emphasises the need for further biological studies and their interaction with hydrological processes as well as temperature in the context of climate change and pandemic scenarios where maintenance could be limited.

4. Discussion

The discussions for this study include descriptions of the limitations faced by this research, as well as the main discussions identified from the results and the future research directions, recommended to fully depict the hydrological performance of vegetated swales under Cfb climate conditions and their impact on a wider SuDS treatment train.

4.1. Limitations of the Study

The investigation presented in this paper evolved under a series of limitations which are relevant to the discussion of the results obtained. The main limitation refers to site conditions, due to the lack of information about the design and construction of the vegetated swale, which forced the research team to undertake forensic engineering to unveil on-site characteristics.

Secondly, the equipment available to develop this research, which started in 2016 with a pressure sensor to measure the water level at the discharge point, only being enhanced in early 2019 with the purchase of a weather station installed in Ryton. These issues limited the investigation to reconstructing weather data for the first 31 months by utilising nearby weather stations from the UK Met Office network. In addition, the research did not have flow meters available to be installed in the inlet of the downspout into the swale to measure the volumes entering the swale, losing the opportunity to carry out a wider study considering the whole treatment train. This problem could be solved in the future through modelling and calibration, which is under development.

Finally, this study defines the long-term hydrological and temperature performance of a vegetated swale under a Cfb climate and a mixed rural and peri-urban environment. This study can be compared to those developed across the world under the same conditions.

4.2. Main Discussion and Future Research Directions

Once the limitations have been described as a framework for the results, the discussion is provided below, also allowing for the identification of future lines of investigation.

- Development of long-term hydrological and temperature models, through modelling and calibration, as added value for the results obtained so far in this descriptive long-term analysis. This would allow researchers and practitioners working under Cfb climates to organise monitoring, operation and maintenance activities through the operational life of the system. This future research line also contributes to strengthening the findings from McCarthy et al. [13] and Lisenbee et al. [29], regarding the minimum number of storm events needed to deliver reliable monitoring data.
- Laboratory based modelling of the swale with scenarios of recent maintenance. This aspect would help to feed modelling parameters where field study is not possible, improving the quality of the predictive models based on real data from the field.
- Obtaining biological models which would allow for the implementation of biological evolution such as vegetation growth and plant science parameters and analyses through modelling, coupling this data to that recorded on-site for this study, which is in line with the findings from Gavrić et al. [16].
- Implement evapotranspiration measures, studying its behaviour depending on the biological evolution and maintenance operations of the vegetated swale.
- This study develops a new research line on the potential design of vegetated swales housing GSHP elements, as this field study describes the long-term temperature performance of a vegetated swale with temporary ponding, being similar to the wet swale design studied in the laboratory by Rey-Mahía et al. [15] which included wet swales with a permanent water level.
- This study will continue to produce data in the following years, becoming one of the few genuine long-term studies developed in the field for the understanding of vegetated swales and their connectivity to other SuDS elements, as well as from a biological evolution stand-point.
- Develop this study under different climatic areas and different environments from rural environments such as transportation infrastructure and consolidated urban environments.

This study is one of the few investigations developed in the long-term under these particular conditions—alongside Andrés-Valeri et al. [33].

5. Conclusions

This study reinforces the need for further long-term monitoring of the hydrological performance of vegetated swales in the field. There is a relevant knowledge gap related to the understanding of the biological evolution of such systems under varying scenarios of operation and maintenance. The work presented in this paper demonstrates the evolution of the system from a simple grassed-vegetated swale into a system that could be described as an emerging wetland, under a scenario with no maintenance activities over a year. This is relevant to scenarios in a changing climate, but also during potential lockdown scenarios, such as that faced during COVID-19, where maintenance activities are limited and non-controlled vegetation growth, as well as reduced maintenance budgets, could be a common scenario.

This study represents one of the largest studies presenting data from three complete hydrological years and 19 storm events captured for detailed analysis during the 6 months stage where on-site weather data were available. Therefore, resulting in a longer period than other significant studies recently published such as those by Leroy et al. [34], Purvis et al. [12], Lisenbee et al. [29] and Andrés-Valeri et al. [33], amongst others.

The ratio to empty temporary water ponding in the vegetated swale ranged from 0.77 mm/h up to 5.00 mm/h for an average of 2.52 mm/h according to stormwater manuals [26] which makes this swale representative as a case study. A total of 19 storm events were identified and isolated in 2019 showing varying storm events with peaks reaching around 25 mm, lasting from hours to several days. The hydraulic performance varied as shown by the water level variations recorded. The monitoring identified the seasonality effect both in precipitation and temperatures, as well as the effect of hot summers; the latter being also highlighted by the previous literature for the specific case of the UK.

Long-term monitoring allows for the identification of recurring storm events which cause potentially hazardous water levels in the vegetated swale, which can activate hydraulic or flooding failures.

Vegetated swales have the capability to incorporate rich biodiversity from the surrounding natural environment in rural areas, such as local plant species and a varied range of pollinator insects, as described in this research under long-term field monitoring.

Vegetated swales serve as thermal regulators, due to their capacity to provide higher surface temperatures than the air temperature in winter and the opposite in summer, providing lower temperatures than the air. This research supports the findings of Rey-Mahía et al. [15] in the laboratory, opening the door for the use of GSHP elements embedded in their cross-section, to heat and cool nearby buildings.

Author Contributions: Conceptualization, L.A.S.-F. and S.J.C.; methodology, L.A.S.-F., J.R.-G., and S.J.C.; software, J.R.-G.; validation, L.A.S.-F. and J.R.-G.; investigation, L.A.S.-F., J.R.-G., S.J.C., E.B.-C., C.R.-M., F.P.Á.-R. and C.L.; resources, L.A.S.-F., S.J.C., J.R.-G. and C.L.; data curation, J.R.-G. and E.B.-C.; writing—original draft preparation, L.A.S.-F. and J.R.-G.; writing—review and editing, S.J.C. and C.L.; supervision, L.A.S.-F., F.P.Á.-R. and S.J.C.; project administration, L.A.S.-F. and F.P.Á.-R.; funding acquisition, L.A.S.-F., S.J.C., and F.P.Á.-R. All authors have read and agreed to the published version of the manuscript.

Funding: This research was funded by Coventry University through the project "Investigation of Green Infrastructure as a combined technique for Bioretention, Flood Resilience and Renewable Energy" and the FICYT through the GRUPIN project, grant number IDI/2018/000221, co-financed with EU FEDER funds.

Acknowledgments: The authors wish to acknowledge the Centre for Agroecology, Water and Resilience (CAWR), Coventry University, for the administrative support, and Garden Organics for allowing the access to the whole experimental site. The authors acknowledge the information provided by the National Meteorological Library and Archive, Met Office, UK (© Crown Copyright 2020). Thanks to Tempcon Instrumental Ltd. and OTT Hydrometry Ltd. for supplying the monitoring equipment. Thanks to Anne-Marie McLaughlin and Milena Tulencic who helped in the installation of the pressure probe and the cleaning of the discharge point of the swale back in 2016.

Conflicts of Interest: The authors declare no conflict of interest. The funders had no role in the design of the study; in the collection, analyses, or interpretation of data; in the writing of the manuscript, or in the decision to publish the results.

References

1. Fletcher, T.D.; Shuster, W.; Hunt, W.F.; Ashley, R.; Butler, D.; Arthur, S.; Trowsdale, S.; Barraud, S.; Semadeni-Davies, A.; Bertrand-Krajewski, J.L.; et al. SUDS, LID, BMPs, WSUD and more—The evolution and application of terminology surrounding urban drainage. *Urban Water J.* **2015**, *12*, 525–542. [CrossRef]
2. Rezaei, A.R.; Ismail, Z.; Niksokhan, M.H.; Dayarian, M.A.; Ramli, A.H.; Shirazi, S.M. A quantity-quality model to assess the effects of source control stormwater management on hydrology and water quality at the catchment scale. *Water* **2019**, *11*, 1415. [CrossRef]
3. Woznicki, S.A.; Hondula, K.L.; Jarnagin, S.T. Effectiveness of landscape-based green infrastructure for stormwater management in suburban catchments. *Hydrol. Processes* **2018**, *32*, 2346–2361. [CrossRef]
4. Shafique, M.; Kim, R.; Kyung-Ho, K. Evaluating the capability of grass swale for the rainfall runoff reduction from an Urban parking lot, Seoul, Korea. *Int. J. Environ. Res. Public Health* **2018**, *15*, 537. [CrossRef]
5. Young, B.N.; Hathaway, J.M.; Lisenbee, W.A.; He, Q. Assessing the runoffreduction potential of highway swales and WinSLAMM as a predictive tool. *Sustainability* **2018**, *10*, 2871. [CrossRef]
6. Avery, L.M. *Rural Sustainable Drainage Systems (RSuDS)*; Environment Agency: Bristol, UK, 2012; ISBN 9781849112772.
7. Ballard, B.W.; Wilson, S.; Udale-Clarke, H.; Illman, S.; Scott, T.; Ashley, R.; Kellagher, R. *The SuDS Manual*; North Carolina State University CIRIA, Griffin Court: London, UK, 2015; p. 968.
8. Abrahams, J.; Coupe, S.; Sañudo-Fontaneda, L.; Schmutz, U. The Brookside Farm Wetland Ecosystem Treatment (WET) System: A Low-Energy Methodology for Sewage Purification, Biomass Production (Yield), Flood Resilience and Biodiversity Enhancement. *Sustainability* **2017**, *9*, 147. [CrossRef]
9. Winston, R.; Hunt, W.F.; Kennedy, S.; Wright, J. *Evaluation of Permeable Friction Course (PFC), Roadside Filter Strips, Dry Swales, and Wetland Swales for Treatment of Highway Stormwater Runoff*; North Carolina State University: Raleigh, NC, USA, 2011; p. 110.
10. Fardel, A.; Peyneau, P.-E.; Béchet, B.; Lakel, A.; Rodriguez, F. Analysis of swale factors implicated in pollutant removal efficiency using a swale database. *Environ. Sci. Pollut. Res.* **2019**, *26*, 1287–1302. [CrossRef]
11. Hunt, W.F.; Fassman, E.A.; Winston, R.J. *NCSU Urban Waterways Factsheet: Designing Dry Swales for the Water Quality Event*; 2016, Extension Fact Sheets, Guidebooks & Design Bulletins - Urban Waterways Series (Stormwater BMP Design Bulletins)' website. Available online: https://stormwater.bae.ncsu.edu/resources/ (accessed on 13 September 2020).
12. Purvis, R.A.; Winston, R.J.; Hunt, W.F.; Lipscomb, B.; Narayanaswamy, K.; McDaniel, A.; Lauffer, M.S.; Libes, S. Evaluating the hydrologic benefits of a bioswale in Brunswick County, North Carolina (NC), USA. *Water* **2019**, *11*, 1291. [CrossRef]
13. McCarthy, D.T.; Zhang, K.; Westerlund, C.; Viklander, M.; Bertrand-Krajewski, J.-L.; Fletcher, T.D.; Deletic, A. Assessment of sampling strategies for estimation of site mean concentrations of stormwater pollutants. *Water Res.* **2018**, *129*, 297–304. [CrossRef]
14. Charlesworth, S.M.; Faraj-Llyod, A.S.; Coupe, S.J. Renewable energy combined with sustainable drainage: Ground source heat and pervious paving. *Renew. Sustain. Energy Rev.* **2017**, *68*, 912–919. [CrossRef]
15. Rey-Mahía, C.; Sañudo-Fontaneda, L.A.; Andrés-Valeri, V.C.; álvarez-Rabanal, F.P.; Coupe, S.J.; Roces-García, J. Evaluating the thermal performance ofwet swales housing ground source heat pump elements through laboratory modelling. *Sustainability* **2019**, *11*, 3118. [CrossRef]
16. Gavrić, S.; Leonhardt, G.; Marsalek, J.; Viklander, M. Processes improving urban stormwater quality in grass swales and filter strips: A review of research findings. *Sci. Total Environ.* **2019**, *669*, 431–447. [CrossRef] [PubMed]
17. Williams, J.B.; Jose, R.; Moobela, C.; Hutchinson, D.J.; Wise, R.; Gaterell, M. Residents' perceptions of sustainable drainage systems as highly functional blue green infrastructure. *Landsc. Urban Plan.* **2019**, *190*, 103610. [CrossRef]
18. Lähde, E.; Khadka, A.; Tahvonen, O.; Kokkonen, T. Can we really have it all?—Designing multifunctionality with sustainable urban drainage system elements. *Sustainability* **2019**, *11*, 1854. [CrossRef]

19. Ariza, S.L.J.; Martínez, J.A.; Muñoz, A.F.; Quijano, J.P.; Rodríguez, J.P.; Camacho, L.A.; Díaz-Granados, M. A multicriteria planning framework to locate and select sustainable urban drainage systems (SUDS) in consolidated urban areas. *Sustainability* **2019**, *11*, 2312. [CrossRef]

20. UK Met Office. UK Climate Averages. Coventry (West Midlands Conurbation). Available online: https://www.metoffice.gov.uk/research/climate/maps-and-data/uk-climate-averages (accessed on 12 August 2020).

21. McGrath, M. Climate Change: UK's 10 Warmest Years All Occurred since 2002. British Broadcasting Corporation Science Environment. 2019. Available online: https://www.bbc.com/news/science-environment-49167797 (accessed on 13 September 2020).

22. Hou, L.; Wang, Y.; Shen, F.; Lei, M.; Wang, X.; Zhao, X.; Gao, S.; Alhaj, A. Study on Variation of Surface Runoff and Soil Moisture Content in the Subgrade of Permeable Pavement Structure. *Adv. Civ. Eng.* **2020**, *2020*, 8836643.

23. Baryła, A.; Karczmarczyk, A.; Bus, A. Role of substrates used for green roofs in limiting rainwater runoff. *J. Ecol. Eng.* **2018**, *19*, 86–92. [CrossRef]

24. Palermo, S.A.; Turco, M.; Principato, F.; Piro, P. Hydrological effectiveness of an extensive green roof in Mediterranean climate. *Water* **2019**, *11*, 1378. [CrossRef]

25. Hill, J.; Drake, J.; Sleep, B.; Margolis, L. Influences of four extensive green roof design variables on stormwater hydrology. *J. Hydrol. Eng.* **2017**, *22*, 4017019. [CrossRef]

26. Maryland Department of Environment. *Maryland Stormwater Design Manual, Vols. I and II*; Maryland Department Of The Environment Water Management Administration: Washington, DC, USA, 2000.

27. Iowa Department of Natural Resources. *Iowa Storm Water Management Manual*; Iowa Department of Natural Resources: Des Moines, IA, USA, 2009.

28. Chen, T.; Ren, L.; Yuan, F.; Yang, X.; Jiang, S.; Tang, T.; Liu, Y.; Zhao, C.; Zhang, L. Comparison of spatial interpolation schemes for rainfall data and application in hydrological modeling. *Water* **2017**, *9*, 342. [CrossRef]

29. Lisenbee, W.; Hathaway, J.; Negm, L.; Youssef, M.; Winston, R. Enhanced bioretention cell modeling with DRAINMOD-Urban: Moving from water balances to hydrograph production. *J. Hydrol.* **2020**, *582*, 124491. [CrossRef]

30. Horstmeyer, N.; Huber, M.; Drewes, J.E.; Helmreich, B. Evaluation of site-specific factors influencing heavy metal contents in the topsoil of vegetated infiltration swales. *Sci. Total Environ.* **2016**, *560–561*, 19–28. [CrossRef] [PubMed]

31. Johnson, J.P.; Hunt, W.F. A retrospective comparison of water quality treatment in a bioretention cell 16 years following initial analysis. *Sustainability* **2019**, *11*, 1945. [CrossRef]

32. North Carolina Department of Environmental Quality. *C-11. Treatment Swale*; North Carolina Department of Environmental Quality Stormwater BMP Manual; North Carolina Department Environmental Quality: Raleigh, NC, USA, 2017; p. 5.

33. Andrés-Valeri, V.C.; Castro-Fresno, D.; Sañudo-Fontaneda, L.A.; Rodriguez-Hernandez, J. Comparative analysis of the outflow water quality of two sustainable linear drainage systems. *Water Sci. Technol.* **2014**, *70*, 1341–1347. [CrossRef] [PubMed]

34. Leroy, M.-C.; Portet-Koltalo, F.; Legras, M.; Lederf, F.; Moncond'huy, V.; Polaert, I.; Marcotte, S. Performance of vegetated swales for improving road runoff quality in a moderate traffic urban area. *Sci. Total Environ.* **2016**, *566–567*, 113–121. [CrossRef] [PubMed]

Article

Effect of Frequency of Multi-Source Water Supply on Regional Guarantee Rate of Water Use

Shanghong Zhang *, Jiasheng Yang, Zan Xu and Cheng Zhang

Renewable Energy School, North China Electric Power University, Beijing 102206, China
* Correspondence: zhangsh928@126.com; Tel.: +86-10-6177-2405

Received: 23 May 2019; Accepted: 27 June 2019; Published: 29 June 2019

Abstract: Multi-source, combined water supply models play an increasingly important role in solving regional water supply problems. At present, in the area of regional water supply, models are mainly used to study the problem of overall water guarantee rate, and do not take into account the impact of the uncertainty of multi-source water supplies on water supply risk. There is also a lack of research on how changes in multi-source water supplies affect sub-region and sub-user water guarantee rates. To address this knowledge gap, the encounter probability of different frequencies and a refined water resources allocation model of multi-source supplies were used. Using Tianjin as an example, this paper studies the quantitative relationship between the uncertainty of multi-source water inflows and the regional guarantee rate of water use. The objectives of the study are to analyze the changing trend of the water shortage rate and the main body of water supply in each region, and to quantitatively describe the influence of the variation of multi-source water supply on the main body of water supply for users. The results show that under the same requirement of guarantee rate for water use, as the number of water diversion sources increase, the probability of water supply meeting the water use rate increases significantly, and the risk to water supplies decreases. At the same time, suburban areas have a low dependence on external water supplies, while the change in the quantity of external water sources has a great impact on the water supply of the Zhongxinchengqu and Binhaixiqnu areas. The distribution and main body of water supply will change for different water users. Therefore, it is important to ensure a stable supply of external water for maintaining the guarantee rate of regional water use.

Keywords: multi-source combined water supply; optimal allocation of water resources; incoming water uncertainty; guaranteed rate of water use

---

1. Introduction

Because of rapid population growth and rapid urbanization, the demand for water resources has greatly increased [1], while water shortages are becoming increasingly serious because of, inter alia, climate change and water pollution [2–5]. Many regions in the world are facing serious water crises and water conflicts [6,7]. Multi-source, combined water supply modes, based on inter-basin water transfers have become an important means for solving urban water supply problems [8,9]. Many countries have planned and implemented a large number of inter-basin water transfer projects [10–12], such as China's South-to-North Water Diversion project [13–16] and California's State Water Project [17]. Inter-basin water transfers alleviate the uneven distribution of water resources through artificial redistribution, and balance the mismatch between water demand and water resources. However, there are disadvantages as well as advantages in the multi-water supply model approach [18]. On the one hand, inter-basin water transfers provide a new water supply source for the water supply system. Multiple water sources can supplement each other to avoid water shortages caused by insufficient water supply from a single water source in low-flow years, and thus effectively reduce the risks to water supplies [19].

On the other hand, because of the differences in water quality, project scale, and water price of the various external water sources [20,21], the competition between multiple water sources is likely to lead to water conflicts in the water distribution process [22]. Therefore, it is of great importance to study the joint water supply from multiple sources for the rational allocation of water resources, the quantification of water guarantee rates, and the assurance of water supply engineering benefits.

One current aspect of multi-source water supply research is to provide or simulate the water supply from each source, and then carry out water-distribution modeling using an optimized allocation model. For example, Montazar et al. studied the combined use of water in irrigation areas, an integrated soil water balance algorithm was coupled to a non-linear optimization model in order to carry out water allocation planning in complex deficit agricultural water resources systems based on an economic efficiency criterion [23]. Based on an integrated GSFLOW model, Wu et al. studied and optimized the combined use of surface water and groundwater in the Heihe River basin of China to alleviate the water conflict between agriculture and ecosystems [24]. Another aspect of multi-source water supply research is to take the amount of water resources as an uncertain condition to determine the upper and lower limits of available water resources, and then simulate the allocation of water resources. For example, Fu et al. adopted a two-stage, interval, stochastic planning method, introduced a risk preference, and carried out research on water resource allocation optimization, using an uncertain stochastic planning model based on risk values for the Sanjiang Plain [25]. Suo et al. proposed a comprehensive solution method for multi-objective, interval programming using fuzzy linear programming and an interactive two-step method, and applied this method to the case study of uncertain multi-water resource joint scheduling planning in eastern Handan, China [26]. In some studies, the above simulation was carried out using typical schemes for the interaction between multiple water sources in high-flow and low-flow years. For example, Yu et al. researched the optimal allocation of water resources in Tianjin based on two scheduling objectives for a system network topology (social benefit and water supply cost) by establishing a multi-source joint scheduling model of the urban water supply system [27]. To address regional water resources allocation under two correlated hydrological random variables and interval parameters, Chen et al. proposed a couple-based interval-bistochastic programming (CIBSP) method. To demonstrate the applicability, the CIBSP method is applied to the Zhanghe Irrigation District located in Hubei Province, Yangtze River basin of China, to optimally allocate available water resources to the municipality, industry, hydropower, and agriculture [28].

In essence, the optimization of the allocation of water resources is a highly complicated risk decision problem [29–31]. The uncertainty in water supply will directly affect the change of water guarantee rate for different water users, and have a great impact on the results of the water resources allocation. However, there is a lack of research on the impact of the uncertainty of multi-source water inflows, and on the probability of combination encounters of water supply risk. There is a lack of research on the impact of water quantity change from multiple water sources on water guarantee rates and the water supply composition for different regions and users. Therefore, the overall objective of this paper is to combine the encounter probability of different frequencies of multiple water sources and a refined water resources allocation model. The specific objectives are to study the quantitative relationship between the uncertainty of multi-source water supplies and the regional water guarantee rate, clarify the water supply schemes for different sources under specific water guarantee rates, and analyze the changing trend of the water shortage rate of users and the main body of water supply in different regions. In addition, the changing trend in water shortage rates for users and the main body of water supply in each region are analyzed, and the influence of the change in multi-source water supply quantity on the main body of water supply is quantitatively described.

2. The Study Area

Tianjin is situated on the banks of the Haihe River in China. Tianjin has a developed economy and a large population. Water resources are in great demand, and there are large local water supply shortages. Per capita water use is only 124.84 m^3 per annum [32]. This is the smallest per capita water use of all the provinces and cities in China. Severe water shortages have greatly restricted the sustainable development of Tianjin. Before the opening of the middle water supply route of China's South-to-North Water Transfer Project (STNWTP), water sources available to Tianjin included surface water, groundwater, reclaimed water, desalinated seawater, and Luanhe river water. The opening of the STNWTP added an additional water source. A multi-source water supply system has been created. According to a preliminary study by the present research team [33], the problem of engineering water shortage can be solved by building a new water distribution network. Tianjin can be divided spatially into a horizontal and a vertical linear feature, three large areal features, eleven partitions, and four users (Figure 1b). The linear features refer to the two major external water diversion projects, i.e., the middle route of the STNWTP and the Luanhe river to Tianjin project. Two straight lines are used to represent the main water supply lines, and ignore the water quantity adjustment function of the reservoir. The three large areal features are the suburban area, the Zhongxincheng area, and the Binhaixinqu area. The eleven partitions refer to eleven administrative divisions within the three large areal features. The suburban areas include Jixian, Baodi, Wuqing, Jinghai, and Ninghe. The Zhongxincheng area includes Zhongxinchengqu and six districts of the city. The Binhaixinqu area includes Beibu, Xibu, Nanbu, Binhaibei, and Binhainan (Figure 1a). The four users refer to the four types of water users in each administrative division: domestic, industrial, agricultural, and ecological water demand. Because of the large number of water supply sources and complex water pipelines, water conflicts will occur among numerous water users with limited water supplies. The complex system of multi-source combined water supply makes the allocation of water resources in Tianjin extremely difficult. Therefore, the water supply system of Tianjin is a good choice for the current investigation's case study. The combination of different frequency encounters of multiple water sources can better reflect the variation of the water guarantee rate for users in different regions. This can be used to quantitatively describe the impact of the fluctuation of water supply from multiple water sources on the actual water supply.

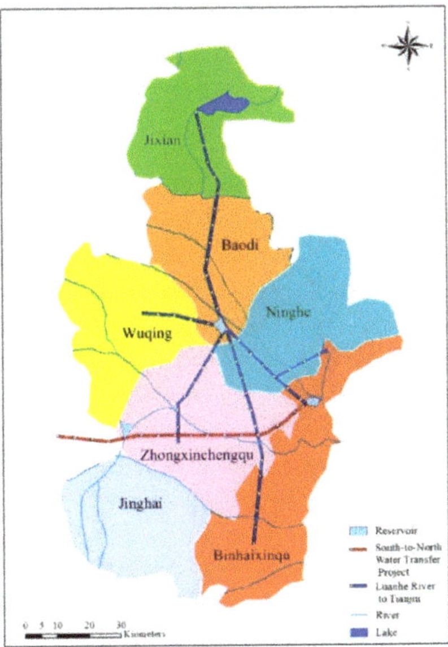

Figure 1. *Cont.*

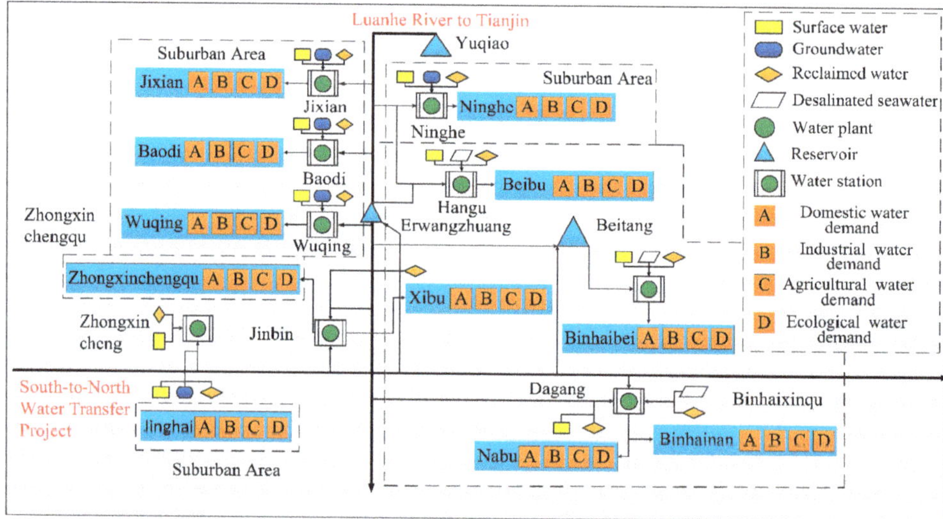

Figure 1. (**a**) Administrative and water distribution map for Tianjin. (**b**) Generalized water supply network. Two straight lines represent the main water supply line of the South-to-North Water Transfer Project and the Luanhe River to Tianjin project in Tianjin. Eleven blocks represent the eleven administrative divisions, and nine "water stations" are connected to two major water supply routes through a network of water pipes. The four users refer to the four types of water users in each administrative division: domestic, industrial, agricultural, and ecological water demand. All available water sources are transferred to the corresponding administrative divisions through the "water station" after unified treatment, and distributed to the four users in need for water.

The water demand data and water supply data for each district in Tianjin in 2030 are shown in Tables 1–3.

Table 1. Water demand forecasting for Tianjin in 2030 (units = 10^8 m^3).

Water Users / Administrative Divisions		Domestic Water Demand	Industrial Water Demand	Agricultural Water Demand	Ecological Water Demand	Total
Zhongxin cheng Area	Zhongxin Chengqu	5.46	3.06	1.17	2.89	12.58
Suburban Area	Jixian	0.46	0.46	0.73	0.05	1.7
	Baodi	0.47	0.18	2.63	0.1	3.38
	Wuqing	0.84	0.3	2.16	0.1	3.4
	Ninghe	0.36	0.37	1.16	0.46	2.35
	Jinghai	0.54	0.28	0.5	0.05	1.37
Binhaixinqu Area	Beibu	0.63	0.3	0.12	0.23	1.28
	Xibu	0.63	0.33	0.12	0.26	1.34
	Nanbu	0.87	1.81	0.12	0.24	3.04
	Binhaibei	1.55	1.52	0.12	0.37	3.56
	Binhainan	0.38	0.82	0.12	0.11	1.43
City		12.19	9.43	8.95	4.86	35.43

Table 2. Prediction units for local water supply in Tianjin in 2030 (units = 10^8 m^3).

Category Administrative Divisions		Surface Water			Groundwater	Reclaimed Water	Desalinated Seawater
		Design Frequency					
		50%	75%	95%			
Zhongxin cheng Area	Zhongxin Chengqu	1.75	1.14	0.56		1.86	
Suburban Area	Jixian	2.24	1.38	0.59	1.88	0.16	
	Baodi	1.02	0.6	0.24	1.11	0.12	
	Wuqing	1.02	0.58	0.21	1.54	0.21	
	Ninghe	0.92	0.57	0.24	0.3	0.13	
	Jinghai	0.9	0.47	0.14	0.32	0.15	
Binhaixinqu Area	Beibu	0.35	0.24	0.13		0.2	1.58
	Xibu					0.21	
	Nanbu	0.66	0.33	0.09		0.56	1.1
	Binhaibei	0.51	0.27	0.09		0.66	
	Binhainan					0.25	0.3
City		9.37	5.58	2.29	5.15	4.51	2.98

Table 3. Water supply from external diversions in 2030 (units = 10^8 m^3).

Category	Design Frequency		
	50%	75%	95%
Luanhe river water		7.5	4.95
STNWTP water	11	10.04	7.24

3. Methods

3.1. Research Design

"Random event", "probability" and "independence" are basic concepts in probability theory. One result of a random experiment is called a random event (abbreviated as event), which is expressed by letters A, B, C, etc. Independent events refer to the fact that the occurrence of one event has no effect on the probability of the occurrence of another event. When event A, B and C are independent of each other, the probability of event A, B and C occurring simultaneously is equal to the product of the probability of three events [34,35]. That is,

$$P(ABC) = P(A) \times P(B) \times P(C) \tag{1}$$

Therefore, the inflow of surface water, Luanhe river water and STNWTP water in Tianjin at different design frequencies are mutually independent events and do not influence each other, i.e., events A, B and C. By calculating the schemes for different combinations of design frequencies of three different water sources, the influence of water quantity changes on guarantee rate of water use of Tianjin is simulated. This could quantitatively describe the distribution of water supplies from multiple water sources to different users in different regions, and describe the influence of the change in water supply sources on the regional water supply.

The flow chart of the research is shown in Figure 2.

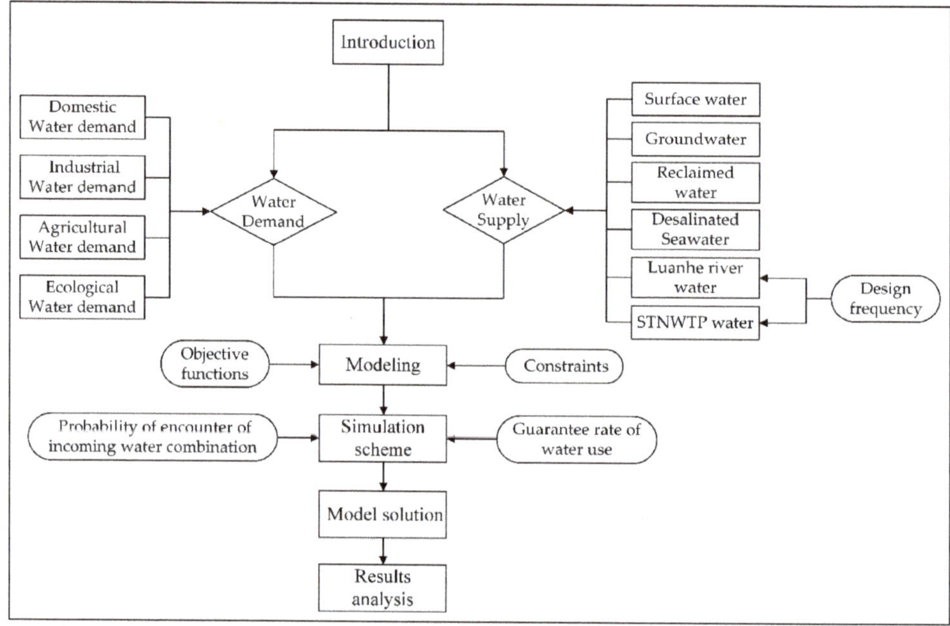

Figure 2. The flow chart of the research.

3.2. Probability of Encounter of Incoming Water Combination and Guarantee Rate of Water Use

After the opening of the middle water supply route of China's STNWTP, water sources available to Tianjin included surface water, groundwater, reclaimed water, desalinated seawater, Luanhe river water, and STNWTP water. Of these, the water supplies from surface water, Luanhe river water, and STNWTP water changed significantly under different design frequencies, while the water supplies from groundwater, reclaimed water, and desalinated seawater was relatively stable. Therefore, the selection of different design frequency combinations for surface water, Luanhe river water, and STNWTP water can effectively reflect the impact on the guarantee rate of water use in different administrative divisions and for different water users. Using the concept of mutually independent events from probability theory [34,35], the inflow of surface water, Luanhe river water and STNWTP water at different design frequencies (50%, 75% and 95%) are mutually independent events and do not influence each other. Assuming that the surface water inflow is event A, the Luanhe river water inflow is event B, and the STNWTP water inflow is event C, the probability that the surface water, the Luanhe river water, and STNWTP water will occur simultaneously at different design frequencies is:

$$P\left(A_{[i]}B_{[j]}C_{[k]}\right) = P\left(A_{[i]}\right) \times P\left(B_{[j]}\right) \times P\left(C_{[k]}\right) \tag{2}$$

where $P\left(A_{[i]}B_{[j]}C_{[k]}\right)$ represents the encounter probability of the three water source combinations. $P\left(A_{[i]}\right)$ represents the surface water when the design frequency is i, $P\left(B_{[j]}\right)$ represents the Luanhe river water when the design frequency is j, and $P\left(C_{[k]}\right)$ represents the STNWTP water when the design frequency is k.

The guaranteed rate of water use is mainly applied to water users, and can be expressed as the degree of water demand by the water users [36]. The higher the guaranteed rate of water use, the higher the satisfaction of the water demand of water users, and the higher the guaranteed water use. The calculation formula is as follows:

$$P = W_s/W_r \tag{3}$$

where W_s represents water supply (1×10^8 m^3), and W_r represents water demand (1×10^8 m^3).

3.3. Construction of Water Resources Optimal Allocation Model

3.3.1. Objective Functions

The benefits of water resources are comprehensive; from the perspective of water resource use, it can be divided into economic benefits and social benefits. The purpose of the optimal allocation of water resources is to ensure that the regional water supply system meets the water demand of all users as far as possible, and minimizes the water shortage rate and achieves the optimal comprehensive benefits of the water resources. Therefore, this paper takes the economic and social benefit targets as the objective functions of the refined water resources allocation model for Tianjin.

Economic Benefit Target

The economic benefit value is calculated in the form of a water benefit coefficient for different users. This can more directly reflect the total benefit value of the water distribution scheme. The economic benefit function is expressed as follows:

$$f_1 x = \max \sum_{i=1}^{I} \sum_{j=1}^{J} \sum_{k=1}^{K} \sum_{l=1}^{L} QG_{i,j,kl} \times a_l - 0.91 \times Q_L - 2.16 \times Q_J \tag{4}$$

where: $f_1(x)$ represents the economic benefits after the completion of the water supply tasks in the various districts during the year (1×10^8 CNY); $QG_{i,j,kl}$ represents the annual amount of water that the water source i supplied through the water station j to the user l of administrative division k (1×10^8 m^3); a_l represents the benefit coefficient of unit water supply to l user (CNY/m^3); Q_L represents the accumulated amount of Luanhe river water (1×10^8 m^3); and Q_J represents the accumulated amount of STNWTP water (1×10^8 m^3).

The industrial benefit coefficient was allocated using the gross industrial output value method [37,38]. This coefficient is the reciprocal of ten thousand CNY of industrial output. The agricultural benefit coefficient is the ratio of agricultural output value to agricultural water consumption. Based on the policy of giving priority to domestic water use to meet and protect ecological and environmental health, the coefficient of living and ecological benefits is taken as a larger value. Details are shown in Table 4:

Table 4. Economic benefit coefficients (units = CNY/m^3).

Category	Calculation Formula
Industrial benefit coefficient	$a_1 = 1/7.65 = 1307$
Domestic benefit coefficient	$a_2 = 1500$
Agricultural benefit coefficient	$a_3 = 467.44/12.32 = 37.95$
Ecological benefit coefficient	$a_4 = 1300$

Where the CNY is the unit of money used in the People's Republic of China.

Social Benefit Target

The social benefit target is to minimize the total water shortage in the region. The social benefits of water resources are manifested in four main aspects: life, industry, agriculture, and ecology. Meeting the domestic water use need is a basic condition for human survival and development. Meeting the industrial and agricultural water use need is a fundamental prerequisite for social stability and continuous economic improvement [39]. Therefore, to maximize the social benefits of water resources when water resources are limited, the minimum total water shortage in Tianjin is selected as the goal of social benefits. The function expression is as follows:

$$f_2(x) = \min\left(\sum_{m=1}^{M}\sum_{n=1}^{N}QX_{m,n} - \sum_{i=1}^{I}\sum_{j=1}^{J}\sum_{k=1}^{K}\sum_{l=1}^{L}QG_{i,j,kl}\right) \tag{5}$$

where $f_2(x)$ represents the water shortage after the completion of water supply allocations in each district during the year (1×10^8 m^3). $QX_{m,n}$ represents the annual water requirements of user n in administrative division m (10^8 m^3). $QG_{i,j,kl}$ represents the annual amount of water that the water source i supplied through the water station j to the user l in administrative division k (1×10^8 m^3).

3.3.2. Constraints

The constraints to the optimal allocation of regional water resources are mainly the constraints of: water supply, water transmission capacity of the pipeline network, and the water supply capacity of the water stations.

(1) Water supply capacity constraints: The total amount of water supplied annually to the connected water stations by the i source during the year was not greater than the maximum capacity of the source.

$$\sum_{i=1}^{I}\sum_{j=1}^{J}QG_{i,j} \leq QG_{i,j_{max}} \tag{6}$$

(2) Pipe network capacity constraints: The amount of water supplied by the i source to the j water station during a year was not greater than the maximum pipeline capacity of the water supplied by the source to the water station.

$$\sum_{i=1}^{I}\sum_{j=1}^{J}QS_{i,j} \leq QS_{i,j_{max}} \tag{7}$$

(3) Water station purification capacity constraints: During a year, the total amount of purified water supplied to each partition user by the j water station was not greater than the maximum capacity of the water purification station.

$$\sum_{i=1}^{I}\sum_{j=1}^{J}\sum_{k=1}^{K}\sum_{l=1}^{L}QG_{i,j,kl} \leq QJ_{i,j_{max}} \tag{8}$$

(4) Water station constraints: The total amount of water supplied by the j water station to users in each district must not be greater than the sum of water supplied by each water source to this water station.

$$\sum_{i=1}^{I}\sum_{j=1}^{J}\sum_{k=1}^{K}\sum_{l=1}^{L}QG_{i,j,kl} \leq \sum_{1}^{I}\sum_{1}^{J}QG_{i,j} \tag{9}$$

(5) Water supply constraints: The sum of the amount of water supplied by the water station to the users in each zone shall not be less than the lower limit of the water demand of the users and shall not be greater than the maximum water demand of the users.

$$\alpha_n * \sum_{i=1}^{I}\sum_{j=1}^{J}\sum_{k=1}^{K}\sum_{l=1}^{L}QX_{i,j,kl} \leq \sum_{i=1}^{I}\sum_{j=1}^{J}\sum_{k=1}^{K}\sum_{l=1}^{L}QG_{i,j,kl} \leq \sum_{i=1}^{I}\sum_{j=1}^{J}\sum_{k=1}^{K}\sum_{l=1}^{L}QX_{i,j,kl} \tag{10}$$

(6) Non-negative constraints: The model satisfies the non-negative constraint of the decision variables.

$$QG_{i,j,kl} \geq 0 \tag{11}$$

3.4. Simulation Scheme Setting

The water resource allocation of Tianjin in 2030 is used as a case study. In this paper, it was assumed that supplies of groundwater, reclaimed water, and desalinated seawater were constant. To analyze the quantitative relationship between the uncertainty of incoming water and the water assurance rate of users in different zones, 33 simulation schemes with different design frequency combinations of surface water, Luanhe river water, and STNWTP water were set up. The detailed simulation scheme settings are shown in Table 5.

Table 5. Simulation scheme settings (units = 10^8 m^3).

Category	Only use Luanhe river water		Only use STNWTP water		
	Luanhe river water		STNWTP water		
Surface water	Design frequency				
	75%	95%	50%	75%	95%
50%	(9.37,7.5)	(9.37,4.95)	(9.37,11)	(9.37,10.04)	(9.37,7.24)
75%	(5.58,7.5)	(5.58,4.95)	(5.58,11)	(5.58,10.04)	(5.58,7.24)
95%	(2.29,7.5)	(2.29,4.95)	(2.29,11)	(2.29,10.04)	(2.29,7.24)

Category	Use both Luanhe river water and STNWTP water		
		STNWTP water	
Surface water	Luanhe river water	Design frequency	
		50%	75%	95%
50%	75%	(9.37,7.5,11)	(9.37,7.5,10.04)	(9.37,7.5,7.24)
	95%	(9.37,4.95,11)	(9.37,4.95,10.04)	(9.37,4.95,7.24)
75%	75%	(5.58,7.5,11)	(5.58,7.5,10.04)	(5.58,7.5,7.24)
	95%	(5.58,4.95,11)	(5.58,4.95,10.04)	(5.58,4.95,7.24)
95%	75%	(2.29,7.5,11)	(5.58,7.5,10.04)	(5.58,7.5,7.24)
	95%	(2.29,4.95,11)	(5.58,4.95,10.04)	(5.58,4.95,7.24)

3.5. Model Solution

LINGO software (Lindo Systems, Chicago, USA) was used to solve the model. LINGO software has a built-in modeling language, which provides multiple internal functions. The overall efficiency of this software, from modeling to solving, is very high. The objective function data and constraint data (water supply source data, user water demand data, water supply network capacity data, daily water plant capacity data, etc.) are input in the form of formulas. The modeling language built into LINGO then automatically transforms the mathematical model into matrix form, and automatically selects the appropriate solver to solve the problem in terms of the model. So LINGO is chosen to solve the model.

4. Results Analysis

4.1. Analysis of the Total Water Guarantee Rate of Different Schemes

4.1.1. Analysis of Six Schemes When the Luanhe River Is the Only External Water Source

Figure 3 shows the occurrence probability for the six schemes, and the trend of the total water use guarantee rate in Tianjin, when the STNWTP water is not used, and when surface water and Luanhe river water are combined with different design frequencies. As can be seen from the Figure 3, with the increase of the total water supply from surface water and Luanhe river water (7.24→16.87 × 10^8 m^3), the guarantee rate of water use of Tianjin gradually increased (53.32%→73.07%), and the occurrence probability of the corresponding scheme presented a downward trend (90.25%→37.5%). Therefore, without the use of STNWTP water, Tianjin has a large water supply deficit. Relying only on local water sources and Luanhe river water is not sufficient for meeting the overall demand for water. At the same time, the risk to water supplies is high and the stability of the urban water supply system cannot be guaranteed.

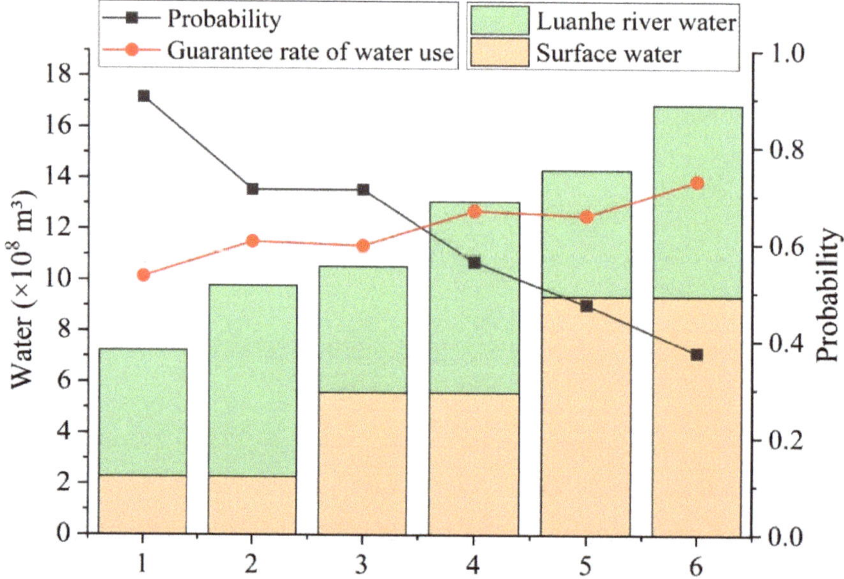

Figure 3. Six schemes when Luanhe river water is the only external water source.

4.1.2. Analysis of Nine Schemes When the STNWTP Is the only External Water Source

To analyze the water supply situation of Tianjin in the absence of Luanhe river water, a comparison was made with the scheme in Figure 3. Nine schemes for surface water and STNWTP water under different design frequency combinations were set up for the case when the Luanhe river water was excluded, as shown in Figure 4. It can be seen from the figure that, the overall guarantee rate of water use for Tianjin is low. As the surface water and STNWTP water supply is increased (9.53→20.37 × 10⁸ m³), the guarantee rate of water use gradually increases (59.78%→82.95%). However, when compared with Figure 3, it can be seen that the minimum and maximum guarantee rates of water use in Figure 4 have been increased, while the occurrence probability of nine schemes has been reduced. By combining the results from Figures 3 and 4, it can be concluded that the water shortages in the aforementioned schemes is caused by insufficient water supplies, which are associated with resource-based water shortages. Water shortages can be alleviated and eliminated by increasing water supply sources.

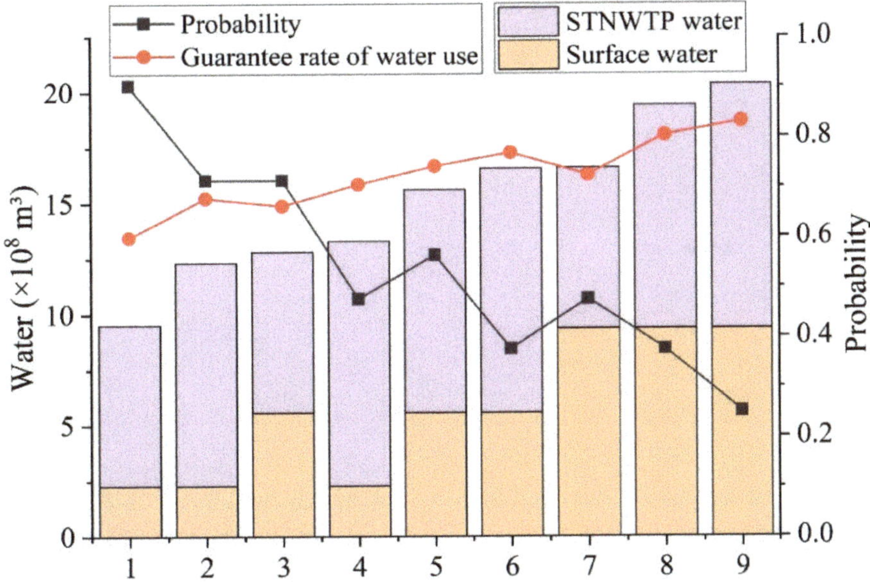

Figure 4. Nine schemes when the STNWTP is the only external water source.

4.1.3. Analysis of Eighteen Schemes When both Luanhe River Water and STNWTP Water Are Used

According to the analysis of the results shown in Figures 3 and 4, there will always be a water shortage in Tianjin when there is only a single external water supply. Therefore, it is necessary to use both Luanhe river water and STNWTP water to increase the amount of water available from external diversions for Tianjin. This will reduce the risk to water supplies and improve the guarantee rate of water use. Figure 5 shows the occurrence probability for eighteen schemes and the variation trend of the overall water use guarantee rate in Tianjin when Luanhe river water and STNWTP water are used simultaneously; different design frequency combinations of surface water, Luanhe river water, and STNWTP water are shown. As can be seen from Figure 5, with the increase in the total water supply of surface water, Luanhe river water, and STNWTP water ($14.48 \rightarrow 27.87 \times 10^8$ m^3), the water guarantee rate of Tianjin gradually increased (73.75% $\rightarrow$ 100%), while the occurrence probability of corresponding schemes showed a downward trend (85.74% $\rightarrow$ 18.75%). This was consistent with the change trends shown in Figures 3 and 4. However, unlike the cases depicted in Figures 3 and 4, in the case of using Luanhe river water and STNWTP water at the same time, the water guarantee rate of all schemes maintained a relatively high level (stable 90% fluctuation), while the minimum and maximum guarantee rates of water use also increased significantly; the minimum water guarantee rate increased to 73.75%, and the maximum water supply guarantee rate increased to 100%.

In summary, with the increase in the types of external water sources, the total water supply in Tianjin increased significantly; the original water shortage has significantly improved, the overall water use guarantee rate shows an upward trend, and can ultimately meet the needs of the users in various urban districts. At the same time, the combination change of different design frequencies of multi-source water leads to uncertainty in water supply, and the fluctuation in water supply greatly affects the water guarantee rate for Tianjin. Therefore, a stable supply of external water sources is of great significance to the urban water supply system of Tianjin.

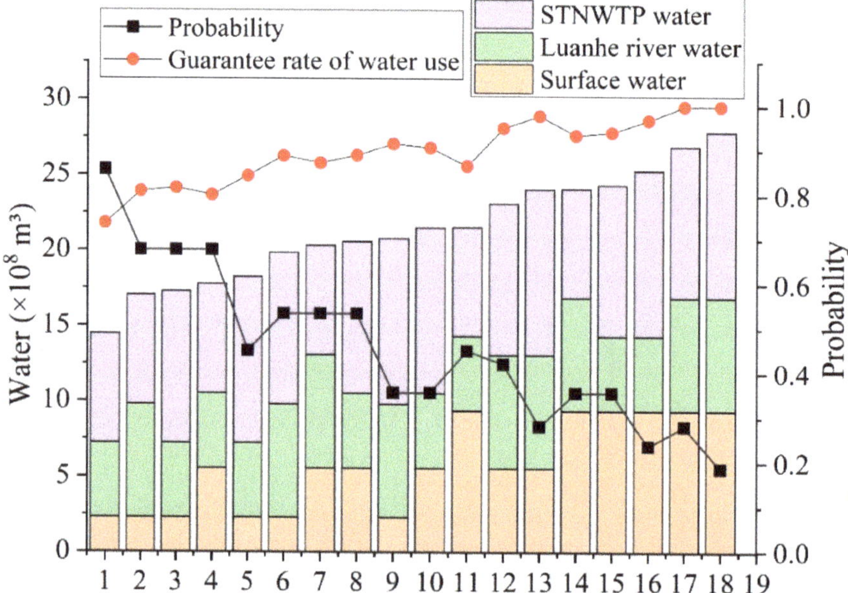

Figure 5. Eighteen schemes when both Luanhe river water and STNWTP water are used.

4.1.4. Occurrence Probability of Water Supply under Different Water Use Guarantee Rate Targets

1. When the water guarantee rate for Tianjin is more than 70%, one of the schemes shown in Figure 3 is satisfied, and the occurrence probability is 37.5%. In Figure 4, it can be seen than six schemes are satisfied, the probability of occurrence is 47.5%, the maximum guaranteed rate of water use is 82.95%, and the probability of occurrence is 25%. In Figure 5, it can be seen that all the schemes are satisfied, the occurrence probability is 85.74%, the maximum water guarantee rate is 100%, and the occurrence probability is 28.13%.

2. When the water guarantee rate for Tianjin is more than 80%, no solution is satisfied in Figure 3. In Figure 4, two schemes are satisfied, and the probability of occurrence is 37.5%. In Figure 5, seventeen schemes are satisfied, and the probability of occurrence is 67.69%.

3. When the water guarantee rate for Tianjin is more than 90%, there are no schemes that are satisfied in Figures 3 and 4. In Figure 5, nine schemes are satisfied, and the probability of occurrence of the schemes is 35.63%.

4. When the water guarantee rate for Tianjin is 100%, only two schemes in Figure 5 are satisfied. Of these, the highest probability is 28.13%.

In summary, from Figures 3–5, the number of schemes that can meet the requirements gradually increases, the occurrence probability of the corresponding schemes with the minimum guaranteed rate of water use shows an upward trend, and the maximum guaranteed rate of water use continuously increases to 100%. This shows that the addition of new water sources, that is, creating a multi-source combined water supply, can effectively improve the guarantee rate of water use, ensure the stability of water supply, and reduce the risk to water supplies. In addition, from a comparison of Figures 3 and 4, it can be seen that the number of water supply schemes in Figure 4 that can satisfy all requirements is higher than that in Figure 3; the corresponding schemes in Figure 4 also have a higher probability of occurrence. Therefore, it can be deduced that the water supply stability of the STNWTP water is better than that of the Luanhe river water, and that the STNWTP is an indispensable water supply source for Tianjin.

4.1.5. The Cost of Water Supply

In a water supply system, the cost of the water supply is an important component. The different water prices of water sources will directly affect the final cost of the water supply. In the above scheme, only the water prices of external water sources (Luanhe river water 0.91 CNY/m^3, STNWTP water 2.16 CNY/m^3) were included in the calculation of the cost of water supply. Only these prices were used, because the purpose was to evaluate the impact of different external water sources on the cost of water supply.

Figures 4 and 5 are taken as examples: the minimum water guarantee rate schemes are compared when the water guarantee rate of Tianjin is more than 70%. The scheme in Figure 4 is: (surface water 95% + STNWTP water 50%), the guaranteed rate of water use is 70.39%, and the cost of water supply is 23.76 × 10^8 CNY. The scheme in Figure 5 is: (surface water 95% + Luanhe river water 95% + STNWTP water 95%), the guaranteed rate of water use is 73.75%, and the cost of water supply is 20.14 × 10^8 CNY. It can be seen that when the guaranteed rate of water use is similar, the water supply cost of the scheme that uses more STNWTP water is higher. On the one hand, the large-scale use of STNWTP water can reduce the risk to water supplies in Tianjin and ensure the stability of water supply system. On the other hand, the high water price of STNWTP water also makes the cost of water supply increase sharply. Therefore, rational allocation of the use of STNWTP water is needed to effectively control the cost of water supply, and to obtain the best economic benefits.

4.2. Analysis of Water Supply to Divisions and Users

To analyze the influence of the variation in the water supplied from the STNWTP on the guaranteed rate of water use in Tianjin, the high-flow years of local water sources and abundant Luanhe river water in Tianjin were selected for analysis. Three schemes in different design frequencies were used in the analysis, design frequency: (surface water 50% + Luanhe river water 75%), (surface water 50% + Luanhe river water 75% + STNWTP water 95%) and (surface water 50% + Luanhe river water 75% + STNWTP water 75%). Based on the water allocation for the three schemes, the variation in water use guarantee rate for different divisions and users in Tianjin were analyzed in order to explore the influence of the variation in STNWTP water supplies on the water supply of users in different divisions.

4.2.1. Analysis of the Overall Water Guarantee Rate for Tianjin

It can be seen from Figure 6 that the overall water guarantee rate for Tianjin clearly increases with the increase in the amount of water that can be supplied by the STNWTP, and that the guarantee rate rises to 100%. Water shortages mainly occur in the agricultural sector. The allocation results show that the reason for this water shortage is that the agricultural water demand in the suburban areas is large, and the local water source cannot meet this demand. Therefore, with the increase in water supply from the STNWTP, the water guarantee rate for the agricultural users gradually increases. As shown in Figure 6a–c, when the water supply from the STNWTP is sufficient, all water requirements can be met.

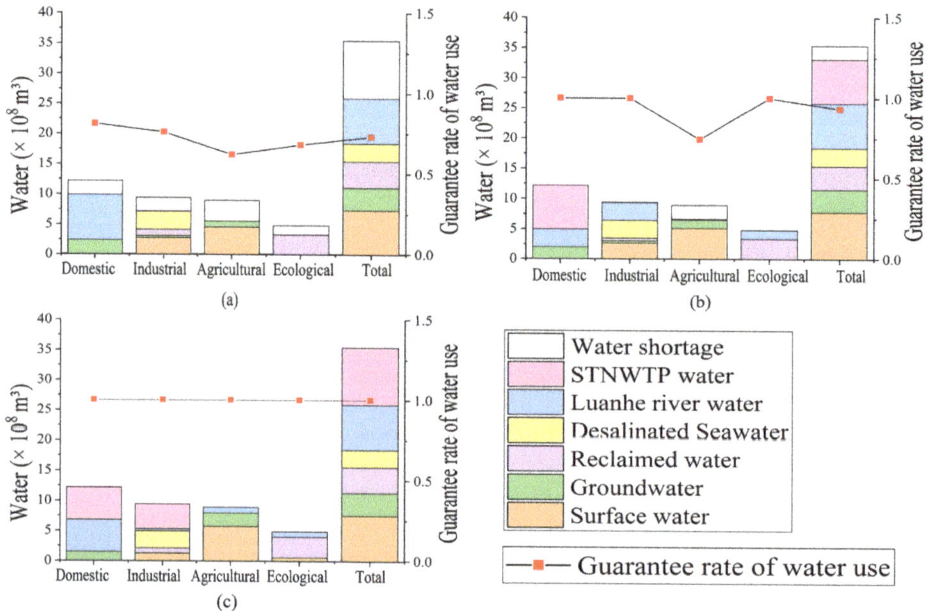

Figure 6. The guarantee rate of water use in Tianjin for: (**a**) (surface water 50% + Luanhe river water 75%); (**b**) (surface water 50% + Luanhe river water 75% + STNWTP water 95%); (**c**) (surface water 50% + Luanhe river water 75% + STNWTP water 75%).

4.2.2. Analysis of Water Supply Guarantee Rate in Suburban Areas

Figure 7 shows the variation in users' water guarantee rate for different schemes in the suburban areas (Jixian, Baodi, Wuqing, Ninghe, and Jinghai). For the scheme that uses Luanhe river water as the only external source (surface water 50% + Luanhe river water 75%, (Figure 7a), the water guarantee rate for the four water users in the suburban areas is generally low, the difference is clear, and there are obvious imbalances in distribution. In the case of the scheme with low amounts of STNWTP water (surface water 50% + Luanhe river water 75%+ STNWTP water 95%, (Figure 7b), the water assurance rate for Wuqing and Ninghe has been significantly improved, and the water distribution between different water users is more equitable. However, the water supply situation in Baodi remained unchanged. The analysis of the results of water distribution show that the water shortages in Baodi mainly occurred in the agricultural sector. The effect of the water distribution rules involving insufficient external water was that the water guarantee rate in the Baodi area did not change. Thus, the conclusion is that the water shortages in the suburban areas in the aforementioned schemes were caused by insufficient water supplies, which can be solved by increasing water supplies. This conclusion is verified in Figure 7c. For the scheme with the highest amount of STNWTP water (surface water 50% + Luanhe river water 75% + STNWTP water 75%), all the water use guarantee rates in the suburban areas are 100%.

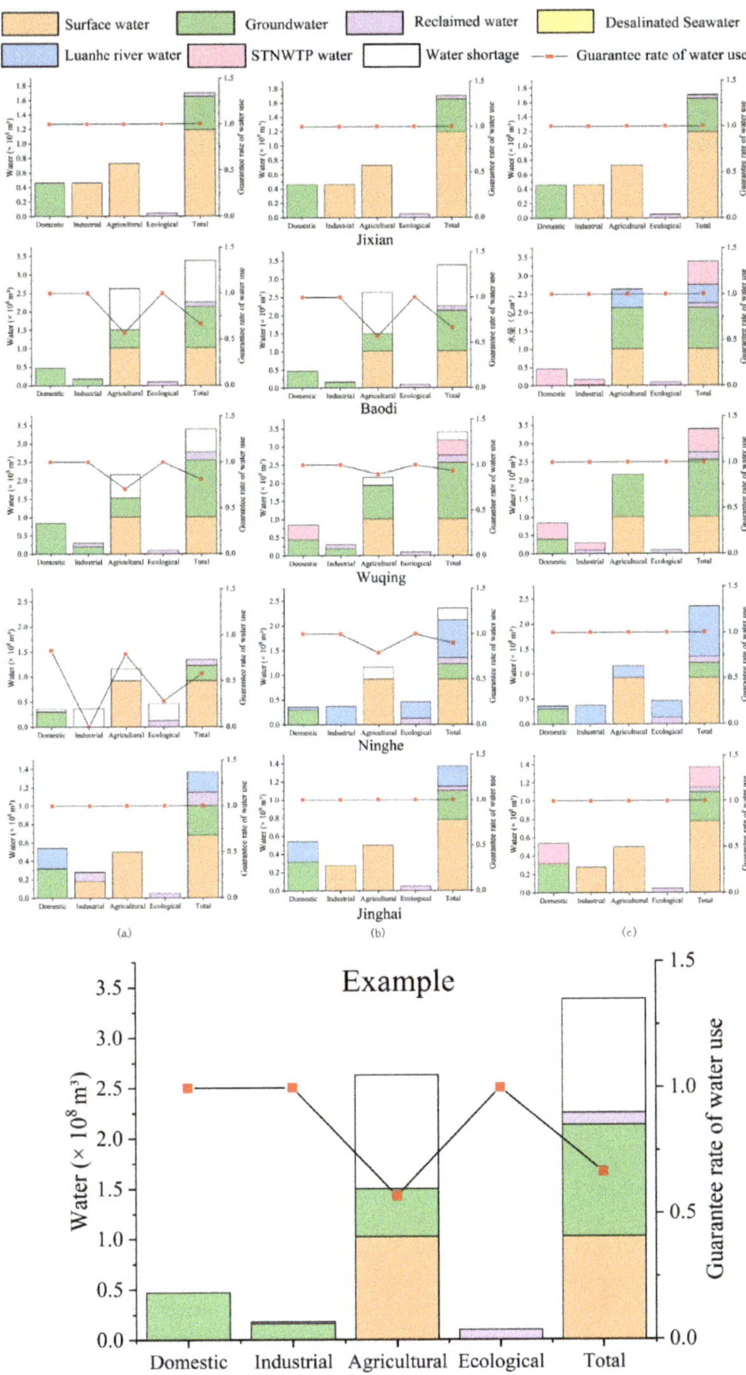

Figure 7. The guarantee rate of water use in suburban areas for: (**a**) (surface water 50% + Luanhe river water 75%); (**b**) (surface water 50% + Luanhe river water 75% + STNWTP water 95%); (**c**) (surface water 50% + Luanhe river water 75% + STNWTP water 75%). All coordinates in the figure are the same and shown in Example.

4.2.3. Distribution of Water Supply in Suburban Areas

The distribution of water supply to various water sources was obtained from the statistics of the water distribution results for users in suburban areas, as shown in Figure 8. As can be seen from the figure, surface water and groundwater in the three schemes are the main source of water supply. The *proportion* of water from the STNWTP increases gradually as the *amount* of water from the STNWTP increases, while the water shortages gradually decrease to zero. It can be seen in Figures 7 and 8 that: (a) groundwater and STNWTP water are the main sources for the domestic and industrial sectors; (b) surface water is the main supply for agricultural water use; and (c) reclaimed water is the main supply for ecological and industrial water requirements. The Luanhe river water changed from supplying water for domestic and industrial water use to supplying water for industrial and agricultural water requirements as the amount of STNWTP water used increased.

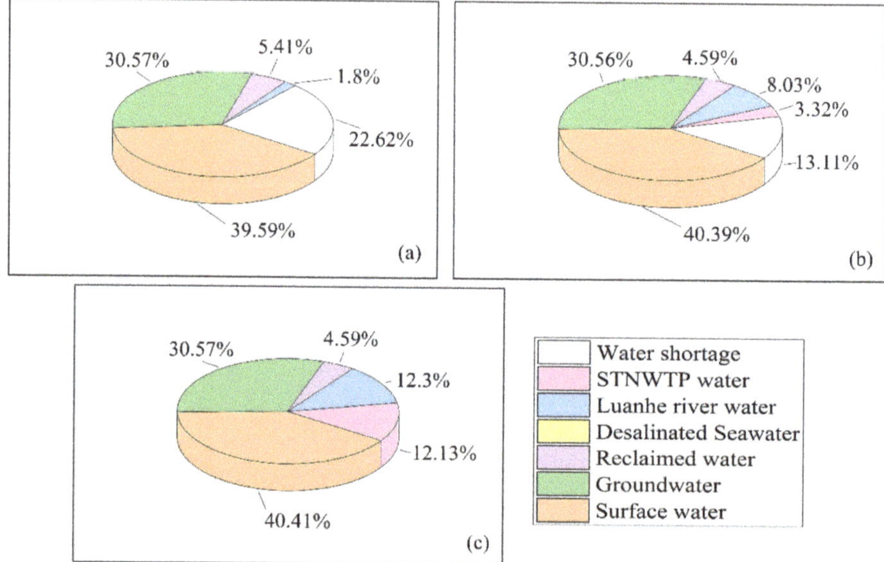

Figure 8. Distribution of water supply in suburban areas for: (**a**) (surface water 50% + Luanhe river water 75%); (**b**) (surface water 50% + Luanhe river water 75% + STNWTP water 95%); (**c**) (surface water 50% + Luanhe river water 75% + STNWTP water 75%).

4.2.4. Analysis of Water Supply Guarantee Rates in the Zhongxinchengqu and Binhaixinqu Areas

Figure 9 shows the variation in users' water guarantee rates for different schemes in the Zhongxinchengqu and Binhaixinqu areas (Zhongxincheng, Beibu, Xibu, Nanbu, Binhaibei, and Binhainan). As can be seen from the figure, as the amount of STNWTP water in the scheme is increased, the water guarantee rate of the four types of water users within the six zones presents an upward trend, until they all of reach 100%. As can be seen in Figure 9a, the shortage of diversion water caused by the lack of STNWTP water leads to serious water shortages in many zones, especially in the water demand from industry and agriculture in Zhongxinchengqu, Xibu, and Binhainan. In the scheme with low amounts of STNWTP water (surface water 50% + Luanhe river water 75% + STNWTP water 95%, (Figure 9b), the water guarantee rate for each region has been significantly improved. Except for a small amount of water shortages for agricultural users in some regions, the water requirements of users can be met. In the scheme with the highest amount of STNWTP water (surface water 50% + Luanhe river water 75% + STNWTP water 75%, (Figure 9c), the water demand requirements of all users within the zones are 100% satisfied.

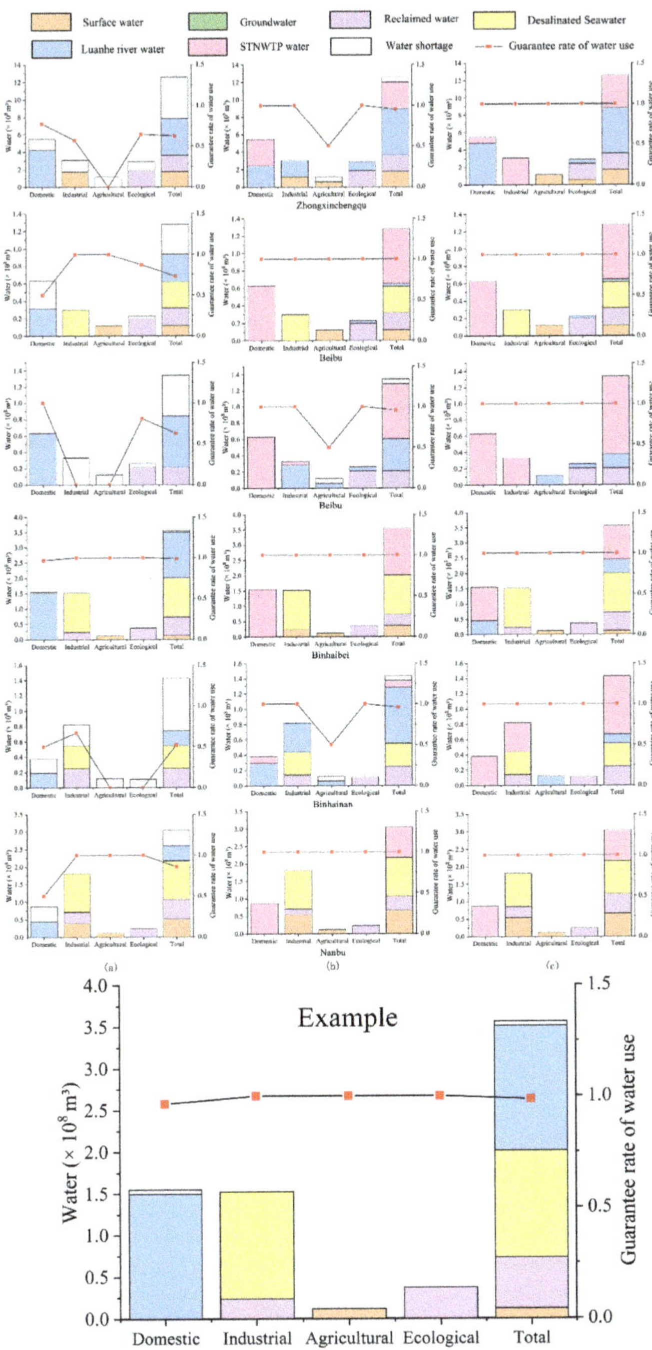

Figure 9. Guarantee rate of water use in the Zhongxinchengqu and Binhaixinqu areas for: (**a**) (surface water 50% + Luanhe river water 75%); (**b**) (surface water 50% + Luanhe river water 75% + STNWTP water 95%); (**c**) (surface water 50% + Luanhe river water 75% + STNWTP water 75%). All coordinates in the figure are the same and shown in Example.

4.2.5. Distribution of Water Supply in the Zhongxinchengqu and Binhaixinqu Areas

Figure 10 shows the water supply distribution for various water sources in the Zhongxinchengqu and Binhaixinqu areas. As the amount of STNWTP water in the three schemes is changed, the distribution of water supply in these areas and the main source of the water supply have changed significantly. In Figure 9a, the highest proportion of water supply in these areas is from Luanhe river water, while in Figure 10b,c, the main water supply is from the STNWTP, and the proportion of STNWTP increases. According to the analysis presented in Figure 9, STNWTP water plays an extremely important role in the water supply system of the Zhongxinchengqu and Binhaixinqu areas. In addition, water in these areas is essentially provided by external diversion water, and thus the areas are highly dependent on external diversion water.

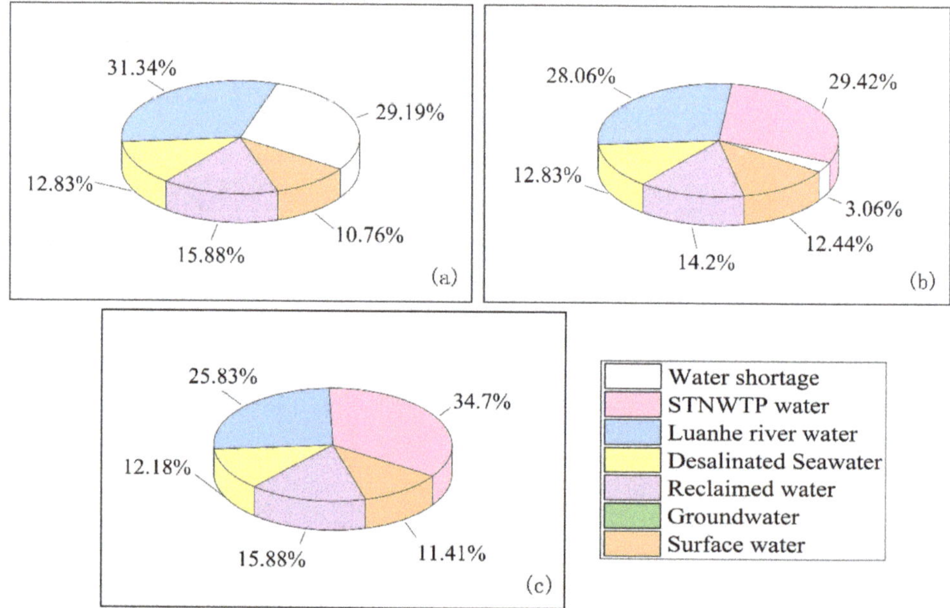

Figure 10. Distribution of water supply in Zhongxinchengqu and Binhaixinqu areas for: (**a**) (surface water 50% + Luanhe river water 75%); (**b**) (surface water 50% + Luanhe river water 75% + STNWTP water 95%); (**c**) (surface water 50% + Luanhe river water 75% + STNWTP water 75%).

In conclusion, water from the STNWTP is one of the most important external water sources for Tianjin, and plays a crucial role in the water supply system of Tianjin. The change in the amount of STNWTP water supplied has a direct impact on the overall water use guarantee rate of Tianjin. By comparing Figures 7 and 9, it can be seen that (a) local water sources (surface water and groundwater) provide a relatively large contribution in the suburban areas, and thus the change in water supply from the STNWTP has little influence on the suburban areas; and (b) the water supply in the Zhongxinchengqu and Binhaixinqu areas mainly depends on external water sources (Luanhe river water and STNWTP water). In addition, STNWTP water accounts for a large proportion of the supply to domestic and industrial users. A change in the amount of water from the STNWTP will have a large impact on the water guarantee rate for the Zhongxinchengqu and Binhaixinqu areas, and will change the water supply distribution and the main source of water for the different water users. Therefore, it is of great significance for the socio-economic development of Tianjin's to: ensure the stability of water supply from the STNWTP, effectively control the fluctuations in water guarantee rate in Tianjin, and maintain the stable operation of the water supply system.

5. Conclusions

The purpose of this investigation was to analyze a regional water supply system that possessed a high degree of complexity. Tianjin was selected for the analysis. The major research design of the paper is using the concept of independent events in probability theory, the investigation combines the encounter probability of different frequencies of multiple water sources with a refined water resources allocation model, it is the major innovation point of the paper. Multiple water sources with different design frequency combinations were set for simulation, and the analysis results can draw the following conclusions.

1. The research analyzes the quantitative relationships between the uncertainties in the multi-source water supplies and the regional water guarantee rate. With the increase in the types of external water sources, the total amount of water supply in Tianjin increases significantly, the guaranteed rate of water use also shows an upward trend. However, the encounter probability of different combinations of design frequencies of multiple water sources gradually decreases.

2. The research reveals the difference of water supply stability of various sources in Tianjin, and the influence of water price of different sources on water supply cost. For different requirements of water guarantee rate, the number of satisfied schemes when using both Luanhe river water and STNWTP water is obviously higher than when a single external source is used. The occurrence probability of the schemes meeting the lowest water guarantee rate clearly improved, and indicates that the water supply stability provided by the STNWTP is higher than that provided by Luanhe river water. However, at the same time, the large dependence on STNWTP water substantially increases the water supply costs.

3. The research describes quantitatively the distribution of water supplies from multiple water sources to different users in different regions, and describes the influence of the change in water supply sources on the main body of water supply for regional water users. Without diversions from the STNWTP water or Luanhe river water, there were serious water shortages for the four water user sectors in Tianjin, especially the industrial, agricultural, and ecological users. The dependence of suburban areas on external water transfers is relatively low, while changes in water quantity from external water sources has a great impact on the water supply to the Zhongxinchengqu and Binhaixinqu areas, and will change the distribution and main source of water supply for different water users.

Author Contributions: Conceptualization, S.Z. and J.Y.; Investigation, C.Z.; Methodology, S.Z., J.Y., Z.X. and C.Z.; Supervision, S.Z. and C.Z.; Validation, S.Z., J.Y. and Z.X.; Writing—original draft, J.Y.

Funding: This study was supported by the 13th Five-Year National Key Research and Development Program of China (2016YFC0401406, 2016YFC0401407) and the Fundamental Research Funds for the Central Universities (2019MS030).

Acknowledgments: The authors thank the Haihe River Water Conservancy Commission for providing the data of water plant and pipeline network, Tianjin Water Bureau for providing the date of surface water, groundwater, Luanhe River water and STNWTP water. We thank Paul Seward, PhD, from Liwen Bianji, Edanz Group China (www.liwenbianji.cn/ac), for editing the English text of a draft of this manuscript.

Conflicts of Interest: The authors declare no conflict of interest.

References

1. Wu, P.L.; Chen, X.H. Sustainable development of Shandong peninsula urban agglomeration: A scenario analysis based on water shortage and water environment changes. *Ecol. Econ.* **2008**, *4*, 189–197.

2. Koleva, M.N.; Calderón, A.J.; Zhang, D.; Styan, C.A.; Papageorgiou, L.G. Integration of environmental aspects in modelling and optimisation of water supply chains. *Sci. Total Environ.* **2018**, *636*, 314–338. [CrossRef] [PubMed]

3. Gebremeskel, G.; Kebede, A. Estimating the effect of climate change on water resources: Integrated use of climate and hydrological models in the Werii watershed of the Tekeze river basin, Northern Ethiopia. *Agric. Nat. Resour.* **2018**, *52*, 195–207. [CrossRef]

4. Ren, C.F.; Li, R.H.; Zhang, L.D. Multiobjective stochastic fractional goal programming model for water resources optimal allocation among industries. *J. Water Resour. Plan. Manag.* **2016**, *142*, 04016036. [CrossRef]

5. Maiolo, M.; Mendino, G.; Pantusa, D. Optimization of drinking water distribution systems in relation to the effects of climate change. *Water* **2017**, *9*, 803. [CrossRef]

6. Huang, G.H.; Chang, N.B. The perspectives of environmental informatics and systems analysis. *J. Environ. Inf.* **2003**, *1*, 1–7. [CrossRef]

7. Wang, L.Z.; Fang, L.P.; Hipel, K.W. Water resources allocation: A cooperative game theoretic approach. *J. Environ. Inf.* **2003**, *2*, 11–22. [CrossRef]

8. Paton, F.L.; Dandy, G.C.; Maier, H.R. Integrated Framework for Assessing urban Water Supply Security of Systems with Non-Traditional Sources under Climate Change. *Environ. Modell. Softw.* **2014**, *60*, 302–319. [CrossRef]

9. Kanakoudis, V.; Tsitsifli, S.; Papadopoulou, A.; Curk, B.C.; Karleusa, B. Water resources vulnerability assessment in the Adriatic Sea region: The case of Corfu Island. *Environ. Sci. Pollut. Res.* **2017**, *24*, 20173–20186. [CrossRef]

10. Tu, X.J.; Du, X.X.; Singh, V.P.; Chen, X.H. Joint risk of interbasin water transfer and impact of the window size of sampling low flows under environmental change. *J. Hydrol.* **2017**, *554*, 1–11. [CrossRef]

11. Juan, L.C. Interbasin water transfers and the size of regions: An economic geography example. *Water Resour. Econ.* **2018**, *21*, 40–54.

12. Annys, S.; Adgo, E.; Ghebreyohannes, T.; Van Passel, S. Impacts of the hydropower-controlled Tana-Beles interbasin water transfer on downstream rural livelihoods (northwest Ethiopia). *J. Hydrol.* **2019**, *569*, 436–448. [CrossRef]

13. Office of the South-to-North Water Diversion Project Construction Committee, State Council, PRC. The south-to-north water diversion project. *Engineering* **2016**, *2*, 265–267. [CrossRef]

14. Huw, P. Compensating for the socioeconomic costs of catchment protection: The case of China's South-North Water Transfer. *Water Secur.* **2019**, *6*, 100023. [CrossRef]

15. Fu, H.; Yang, X.L. Effects of the south-north water diversion project on the water dispatching pattern and ecological environment in the water receiving area: a case study of the fuyang river basin in handan, China. *Water* **2019**, *11*, 845. [CrossRef]

16. Kuo, Y.M.; Liu, W.W.; Zhao, E.M. Water quality variability in the middle and down streams of Han River under the influence of the Middle Route of South-North water diversion project, China. *J. Hydrol.* **2019**, *569*, 218–229. [CrossRef]

17. Lefkoff, L.J.; Kendall, D.R. Optimization modeling of a new facility for the California state water project. *JAWRA J. Am. Water Resour. Assoc.* **1996**, *32*, 451–463. [CrossRef]

18. Huang, W.; Chen, J. Approach on running mechanisms of water right allocation and water market in interbasin diversion project. *J. Yangtze River Sci. Res. Inst.* **2006**, *23*, 50–52. (In Chinese)

19. Puleo, V.; Fontanazza, C.M.; Notaro, V. Multi sources water supply system optimal control: a case study. *Procedia Eng.* **2014**, *89*, 247–254. [CrossRef]

20. Jean-Christophe, P. Conflicting Objectives in Groundwater Management. *Water Resour. Econ.* **2018**, in press. [CrossRef]

21. Lopes, A.F.; Macdonald, J.L.; Quinteiro, P.; Arroja, L.; Carvalho-Santos, C.; Cunha-e-Sá, M.A.; Dias, A.C. Surface vs. Groundwater: The Effect of Forest Cover on the Costs of Drinking Water. *Water Resour. Econ.* **2018**, in press. [CrossRef]

22. Martinsen, G.; Liu, S.; Mo, X.; Bauer-Gottwein, P. Joint optimization of water allocation and water quality management in Haihe River basin. *Sci. Total Environ.* **2018**, *654*, 72–84. [CrossRef] [PubMed]

23. Montazar, A.; Riazi, H.; Behbahani, S.M. Conjunctive water use planning in an irrigation command area. *Water Resour. Manag.* **2010**, *24*, 577–596. [CrossRef]

24. Wu, X.; Yi, Z.; Wu, B. Optimizing conjunctive use of surface water and groundwater for irrigation to address human-nature water conflicts: A surrogate modeling approach. *Agric. Water Manag.* **2016**, *163*, 380–392. [CrossRef]

25. Fu, Q.; Li, J.H.; Liu, D.; Li, T.X. Allocation optimization of water resources based on uncertainty stochastic programming model considering risk value. *Trans. Chin. Soc. Agric. Eng.* **2016**, *32*, 136–144.

26. Suo, M.Q.; Wu, P.F.; Zhou, B. An integrated method for interval multi-objective planning of a water resource system in the eastern part of Handan. *Water* **2017**, *9*, 528. [CrossRef]

27. Yu, B.; Liang, G.H.; He, B. Modeling of joint operation for urban water-supply system with multi-water sources and its application. *Adv. Water Sci.* **2015**, *26*, 874–884. (In Chinese)

28. Chen, S.; Xu, J.J.; Li, Q.Q. A copula-based interval-bistochastic programming method for regional water allocation under uncertainty. *Agric. Water Manag.* **2019**, *217*, 154–164. [CrossRef]

29. Singh, A. Simulation–optimization modeling for conjunctive water use management. *Agric. Water Manag.* **2014**, *141*, 23–29. [CrossRef]

30. Hassan-Esfahani, L.; Torres-Rua, A.; Mckee, M. Assessment of optimal irrigation water allocation for pressurized irrigation system using water balance approach, learning machines, and remotely sensed data. *Agric. Water Manag.* **2015**, *153*, 42–50. [CrossRef]

31. Abdulbaki, D.; Al-Hindi, M.; Yassine, A. An optimisation model for the allocation of water resources. *J. Clean. Product.* **2017**, *164*, 994–1006. [CrossRef]

32. Tianjin Municipal Bureau of Statistics. *Tianjin's Statistical Yearbook*; Tianjin Municipal Bureau of Statistics: Tianjin, China, 2015. (In Chinese)

33. Zhang, S.H.; Yang, J.S. Multi-water source joint scheduling model using a refined water supply network: case study of Tianjin. *Water* **2018**, *10*, 1580. [CrossRef]

34. Michael Steele, J. *Probability Theory: Formal. International Encyclopedia of the Social & Behavioral Sciences*, 2nd ed.; Cambridgle University Press: Cambridge, UK, 2015; pp. 33–36.

35. Ross, M.S. *Introduction to Probability Theory. Introduction to Probability Models*, 12th ed.; Academic Press: Amsterdam, The Netherlands, 2019; pp. 1–21.

36. Hou, B.D.; Gao, E.K.; Xu, Z.Z. Connotation, calculation and application of guarantee rate of water use. *China Water Resour.* **2015**, *17*, 12–15. (In Chinese)

37. Lavric, V.; Pertrica, l. Optimal water system topology through genetic algori-thm under multiple contaminated-water sources constraint. *Comput. Aided Chem. Eng.* **2004**, *18*, 433–438.

38. Wang, L.T. *Detailed Water Allocation of Urban. Complex. Water Utilization System—Case Study in New District of Tianjin*; China Institute of Water Resource and Hydropower Research (IWHR): Beijing, China, 2015. (In Chinese)

39. Xu, Y.; Wang, Y.; Li, S. Stochastic optimization model for water allocation on a watershed scale considering wetland's ecological water requirement. *Ecol. Indicat.* **2018**, *92*, 330–341. [CrossRef]

MDPI
St. Alban-Anlage 66
4052 Basel
Switzerland
Tel. +41 61 683 77 34
Fax +41 61 302 89 18
www.mdpi.com

Water Editorial Office
E-mail: water@mdpi.com
www.mdpi.com/journal/water